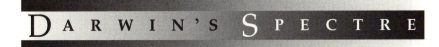

DARWIN'S SPECTRE

DARWIN'S SPECTRE

Evolutionary

Biology in

the Modern World

MICHAEL R. ROSE

PRINCETON UNIVERSITY PRESS • PRINCETON • NEW JERSEY

Published by Princeton University Press, 41 William Street,
Princeton, New Jersey 08540
In the United Kingdom: Princeton University Press, Chichester, West Sussex

Third printing, and first paperback printing, 2000

Paperback ISBN 0-691-05008-2

The Library of Congress has cataloged the cloth edition of this book
as follows:

Rose, Michael R. (Michael Robertson), 1955–
Darwin's spectre : evolutionary biology in the modern world /
Michael R. Rose.
p. cm.
Includes bibliographical references (p.) and index
ISBN 0-691-01217-2 (cloth : alk. paper)
1. Evolution (Biology)—Social aspects. 2. Natural selection—
Social aspects. I. Title.
QH371.R683 1998
576.8'2—dc21   98-11494 CIP

This book has been composed in Palatino

The paper used in this publication meets the minimum
requirements of ANSI/NISO Z39.48-1992 (R1997)
(Permanence of Paper)

http://pup.princeton.edu

Printed in the United States of America

3 4 5 6 7 8 9 10

FOR MY IMPS, *Caitlin and Liam*

Contents

ACKNOWLEDGMENTS

THIS BOOK is the accidental creation of years of teaching, tutoring, and public speaking at four universities in three countries. Begun with no clear purpose in the summer of 1987, it struggled to a conclusion in 1998. Over that expanse of time, I have been counseled and encouraged by many colleagues and friends whom I will not acknowledge here. The vagaries of time and my memory must provide me with my excuses. I am grateful above all to the many students at the University of California, Irvine, whose questions helped me to ripen my material.

Of the specific debts that come to mind, my biggest is to Chris Moore. He played an indispensable role in the development of the ideas and arguments of Part Three. I look forward to a book by Chris, possibly with myself, exploring the psychobiological ramifications of immanent Darwinism. Much that I have left unclear or unfocused will no doubt then be made transparent.

The man most responsible for the publication of this book is Jack Repcheck of Princeton University Press. He has been the producer of this book and has pushed me along, supplied criticisms and encouragements as appropriate, and refused to listen to pleas for delay, reconsideration, or second thoughts.

A special word of thanks for Greg Benford, who has been a mentor to me as a writer. His words of editorial advice, scientific perspective, physics trivia, and gossip about publishing have played a major role in my development as an author.

This book has had many drafts and many readers. Other than the aforementioned, those who gave me useful comments include Roger Gosden, John Knight, George Lauder, Margarida Matos, Roger McWilliams, Nicholas Metal, Randolph Nesse, Theodore Nusbaum, and Jay Phelan. Michael Ruse and Steven Austad reviewed the book for Princeton University Press, providing many illuminating comments. I apologize to those whom I have overlooked. Naturally, I received much advice that I could not follow, due to my own limitations.

Finally, this book was written with the support and faith of Della Rose, whose unwarranted belief in my ability to write was its main inspiration.

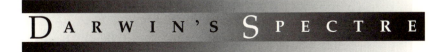

INTRODUCTION

A SPECTRE is haunting the modern world, Darwin's Spectre, Darwinism. This spectre has frightened religious ministers, curdled school curricula, and left the politically correct ill-at-ease. The rhetoric and posturing have been long sustained, from the nineteenth-century clash of Bishop Wilberforce and T. H. Huxley over Darwin's *Origin of Species* to the Scopes monkey trial in Tennessee in 1925 to the Scientific Creationists of the 1980s, who attempted to legislate the teaching of biblical doctrines in juxtaposition to Darwinism. Alone among modern scientific doctrines, Darwinism has upset many beyond the academy. It has had enemies on the right and the left, and some states, such as the Soviet Union, liquidated Darwinians for their scientific beliefs. Yet Darwinism goes on as a scientific movement. It even has its lay devotees, some of whom name their sons Darwin. The centennial of Darwin's death, 1982, brought forth an outpouring of festive celebrations in the towns of Italy, Spain, and Greece. Darwin T-shirts are everywhere. The only competition the Jesus fish gets among metal car fixtures is the Darwin fish with four legs. Darwin is an emblem of resistance to clerics, to the orthodox everywhere. Darwin's Spectre won't go away.

This book is a personal view of the nature of Darwinism and an interpretation of its peculiar haunting of the modern world. It is not a scholarly or historical catalog of all pertinent learned debate, though it mentions many of the connections between Darwinism and other modern ideas. My goal is to present Darwinism, its milieu, and its implications from the perspective of an evolutionary biologist. This is necessarily a subjective view. Particularly with respect to the clash between Darwinians and their persecutors, no evolutionary biologist can claim objectivity. Too often we feel like we are clinging to a mountainside, the winds of controversy blowing around us. This has led many of us into confusion and equivocation. Some of us dissimulate, and pretend that evolution is some benign force in the world. Bromides about evolutionary progress have deformed our popular writings. We are cowards too often, and when not, usually foolhardy.

I need to address some potential misunderstandings. Most importantly, I do not intend to mount a case for edifying effects resulting from Darwinian beliefs. The truth is that Darwinism has been used to bolster such social movements as eugenics, which led to the abuse, sterilization, or death of many innocent people. There is no good argument to be made that Darwinism is an important source of moral or social enlightenment. Admittedly, there have been eras, such as the late Victorian age, when Darwin was seen as a secular saint in the church of human progress. But at that same time, Darwinism was also an explicit buttress for Aryan supremacists. Darwinism has no inherent virtue. Darwin's Spectre is neither an evil ghost, nor an angel. It is instead an ambiguous and troubling apparition, from which we might learn much.

Though it is common to explain the actions of men and women in terms of the will of the gods, as in Homer, or the material infrastructure, as in Marx, I have no firm opinion about the influence of era, technology, or favorable constellations on human beliefs or achievements. In some instances, these effects seem to me to be indubitable; other cases I am uncertain about.

I also do not suppose that the difficulties and confusions of various Darwinists, including Darwin himself, reveal that science is a nonprogressive dinosaur that has been given too much status by governments or universities. Darwinism, like all sciences, has had some dire times, but it eventually recovered. Scientific progress seems to me neither certain nor impossible.

I do not think that Darwinism clearly shows us the whole truth concerning any particular ethical or moral question. But this is not to say that the calculations, inferences, or results of Darwinists are never relevant to such questions. The findings of Darwinism may occasionally be more germane to "social" questions than religious writings that are thousands of years old, among other claimants to wisdom. However, this is not a very high distinction.

On the other hand, Darwinism is highly pertinent to many of the thorniest practical problems of agriculture, medicine, and related fields. In many areas of applied science, Darwinism has been appreciated too little. It is my firm belief that considerable improvements will be achieved across a range of technologies once Darwinism is properly exploited.

With subject matter as broad as one of the leading intellectual movements of modern times, much had to be left out of this book. My foremost dereliction has been no serious discussion of the many manifestations and refractions of Darwinism in literature, where I construe literature rather broadly as ranging from science fiction to existentialist philosophy. This task would have been endless, because the literary impact of Darwinism has been so widespread, but so often incoherent. The misunderstandings of those who write without fiction have been numerous enough. The myriad confusions of those who have rendered Darwinism by literary devices entirely defeat my imagination.

This book is divided into three parts. In Part One, "Darwin and Darwinian Science," I distill the main elements of Darwinism as a scientific project. The story begins with the Darwin family, but the ideas of evolutionary biology soon take over from the vicissitudes of the Darwin brood. These chapters present the main outlines of Darwin's achievement and its improvement by his successors.

Part Two, "Applications of Darwinism," deals with the use of evolutionary thought to address issues of material human concern. These issues range from the prosaic, such as agriculture, to the Promethean, such as the attempt to direct the breeding of the human species. I attempt to show how Darwinian thought is of fundamental importance for the understanding or improvement of each of these undertakings.

The third part of the book, "Understanding Human Nature," considers two different Darwinian approaches to the evolution of man and the human psyche. These two theories are at opposite ends of the spectrum with respect to a variety of questions, including the following. How different is human behavior from that of other animals? Can human behavior be predicted from evolutionary theory? What are the prospects for scientifically based intervention in human affairs, economic or political? What is the source of religious experience? These questions can't be answered definitively at this time, but Darwinian thinking provides some interesting angles from which to view them.

With such a broad agenda, this book necessarily lacks the copious and fine detail that many academics would prefer. I write instead for the general reader, curious about evolution and its meaning, intelligent enough to seek some enlightenment, but not at leisure to read a dozen or more volumes on the subject. Here I offer a large canvas, but the painting is more Van Gogh than Michelangelo in its use of brushstrokes. I hope that the reader's eye is pleased nonetheless.

PART ONE

DARWIN AND DARWINIAN SCIENCE

INTRODUCTION TO PART ONE

THIS FIRST PART sketches Darwinian science, starting with the career of Darwinism's great founder. There are three major issues in evolutionary biology: the nature of heredity, the operation of selection, and the pattern of evolution. Each of these issues has a long intellectual history, predating Darwin. Some of that history will be supplied here, to the extent that it helps clarify the ideas. History for its own sake will not be prominently featured, despite the low regard that historians have for the introduction of historical material couched in terms of present-day significance. Beyond their historical development, I strive to summarize the core thinking about each of the three main themes of Darwinian science. Some of these ideas were originally developed mathematically, but I will do no more than offer heuristic descriptions of that mathematics. Darwinian theory is not intellectually fearsome the way that theoretical physics is. At the same time, Darwinism isn't as marvelously idea-free as most of biology, which tends to pile fact upon fact, with great lashings of terminology. It is more like economics in its quality of being both abstract and intuitive. Finally, some of the key findings and experiments will be described, in order to convey the point that Darwinism is, after all, empirical science, not just metaphysics.

I make no pretense that these sketches are more than my favorite episodes and arguments from the annals of Darwinism. However, they may provide points of entry for further reading. I recommend that the reader skip over passages that do not make sense on first reading. The difficulty may reflect my limitations as a writer. The main points should emerge even if some of the more technical points are not adequately conveyed. The intellectual structure of Darwinism has considerable inner strength and coherence, if not beauty.

1 DARWIN

The Reluctant Revolutionary

DARWINISM is an outgrowth of the mind of Charles Robert Darwin. There is still no experience more bracing for a young evolutionary biologist than reading Darwin's own words, particularly his *Origin of Species*. Thus the starting point for any discussion of Darwinism must be the man himself, as he grew up in Regency England. To some extent, this is a refreshing story, because Darwin was no intimidating prodigy. He was a sound lad growing up in the landed gentry, a better character for a romance novel than science fiction. And yet, there is much in his background that points directly to the great scientific figure he would later become, even to the very substance of his scientific discoveries.

Here the bare narrative of Darwin's life will be supplied, supplemented by a few remarks about his scientific work, conceived broadly. Darwin's scientific work will then be treated in more detail in chapters 2, 3, and 4. However, it should be understood that this entire volume lies within Darwin's shadow. Little of our discussion could be said to lack the distinctive stamp of the man who returned from the voyage of H.M.S. *Beagle* in 1836.

DARWIN'S CULTURAL AND FAMILY HERITAGE

To understand Darwin's background, it is necessary to understand the significance of the Enlightenment of the eighteenth century for British culture. While the Enlightenment in France was characterized by bold and provocative writers, in Britain less colorful figures like David Hume, Edward Gibbon, and Adam Smith laid the foundations for modern philosophy, modern history, and the new field of economics. If there was any single trailblazer in this group, it was Hume.[1] The second son from meager Scottish nobility,

Hume did more to clear the ground for British thought during the Enlightenment than any other person. Key to this role were Hume's anonymous and posthumous arguments against religion. The first of the modern, which is to say skeptical, epistemologists, for Hume nothing was to be taken on faith, especially not Faith itself. There is a story about pre-Revolutionary France, in which the mistress of a Prince of the Blood came upon him reading Hume on his death-bed. She immediately burst into tears, and demanded to know how he could read such heresy so close to his death. The Prince replied, "When you have led the life I've led, this is very comforting." Hume himself frequented the French Court, where his charm earned him the nickname "Le bon David." In fact, he had a famous mistress there, though he shared her with a prince. For some reason, he did not literally lose his head, despite his figurative loss of the same structure. Hume did not even renounce his atheism as he lay dying of cancer in 1776, the year his friend Adam Smith published *The Wealth of Nations* and Thomas Jefferson drafted the Declaration of Independence, both classics of the Enlightenment, at least in the British style.

The importance of thinkers like Hume and Smith to Darwinism was that they developed a style of reasoning which would prove to be fundamental to Darwinian thought. In this style, the writer reduces a situation to its barest elements, an approach akin to reductionism, and then attempts to set these elements "in motion" according to some simple dynamical principles. A classic example of this approach is Adam Smith's analysis of the market in terms of individual manufacturers. Such analyses can be characterized in terms of their clarity, logic, and daring. It should also be noted that these were by no means hallmarks of British thinking at the time. The writings of Voltaire, Casanova, and other "Continental" men and women of letters had the same quality.[2] Indeed, the very speech of eighteenth-century Europe took a turn away from polite obscurity toward provocative clarity: logical, didactic, and incisive. It was truly the temper of the age, and the temper of Charles Darwin's grandfather, Erasmus.

Charles Darwin was one of the greatest intellectual revolutionaries of all time. Surely such a man must have been a fire-breathing, nonconformist, outrage of a human being? Well, as we will see, he wasn't. But his paternal grandfather, Erasmus, was indeed made of cloth that fit this pattern.

Erasmus Darwin (1731–1802) was a leonine figure from profligate times, the eighteenth century. Among his contemporaries were Voltaire, Frederick the Great, Giacomo Casanova, Mozart, and David Hume. Like them, he lived a "big" colorful life. Like them, he also left a trail of personal debris and confusion.

Erasmus was first and foremost a physician. Like his sons and grandsons, Erasmus studied medicine at Edinburgh University, the leading medical school of the day. Despite, or perhaps because of, the low intellectual quality of medicine at that time, Erasmus had an extremely fertile mind. He was full of speculations about engineering as well as both physical and biological science, including schemes for making money from these ideas. This intellectual vitality led Erasmus to form a club for upcoming members of the new scientific and technological elite, the Lunar Society. Among the members were Josiah Wedgwood, James Watt, Benjamin Franklin (who would later become one of America's Founding Fathers), and Joseph Priestley, notable figures in ceramics, engineering, physics, and chemistry, respectively. These connections, together with some chemical experiments, led to the election of Erasmus to the Royal Society, England's premier scientific organization. Erasmus Darwin was one of the first chemists to abandon the phlogiston theory of combustion, now known to be erroneous. Erasmus published epic works on biology and geology, including *Zoonomia* and *Phytologia*, which incorporated proto-evolutionary speculations, though of an ill-formed kind. These works were quite influential at the time, and sold well. Mary Shelley acknowledged the inspiration of this Dr. Darwin in the Preface to her novel *Frankenstein*.[3]

In all this activity, Erasmus Darwin epitomized the helter-skelter creativity of many of the leading minds of the eighteenth century. This Enlightenment was a Promethean time when, it was then thought, the powers of Reason would lay bare the secrets and possibilities of Nature. This was an epoch when long-standing tradition was seen primarily as an impediment, a time when people began openly speculating about entirely secular societies and the overthrow of religion or monarchy. And of course many of these things were actually happening in places like the new United States of America and the first Republic of France. If history were logical, then this would have been the time when a full-blown theory of evolution would have been developed and published, most likely by Erasmus Darwin. As history is not a well-crafted novel, but "a tale told by an idiot," Erasmus Darwin did not found evolutionary biology. That would be left to a grandson he never saw.

One final anecdote: The intermittent madness of King George III due to porphyria, an inherited metabolic disorder, led to desperate attempts to find a suitable physician for his condition, in effect a psychiatrist. Something of this desperation is conveyed in the film *The Madness of King George*. Erasmus Darwin was such an eminent physician at the time, and so well known for his advanced under-standing of the human mind, that he was offered the post of the King's personal physician, a job that would inevitably have led him to be installed in the aristocracy. But Erasmus turned the offer down.

Darwin's other grandfather was also famous: Josiah Wedgwood (1730–95). His family's trade was pottery, at which he worked from the age of nine. In his twenties he became dissatisfied with the poor quality pottery that was then produced in England, and he at-tempted to develop the field of ceramics by a course of extensive experimentation. One of his inventions was an improved type of earthenware, called "creamware," which sold quite well, even to the royal family. This led to the rapid expansion of his business and the beginning of the Wedgwood fortune. Josiah Wedgwood was a self-made captain of industry, a man of little formal education, with a great instinct for self-improvement and the improvement of man-ufactures. Nonetheless, Wedgwood was a nonconformist and radi-cal like Erasmus Darwin, always searching for new ways to under-stand the world by the effort of his own mind, rather than authority. Wedgwood was initially one of Darwin's patients. Friendship between the Wedgwoods and Darwins began over the funding of the Grand Trunk canal, for which Josiah Wedgwood sought Erasmus Darwin's help. The ten years it took to fund and build the canal cemented an alliance between the two families which was to last for more than a century.

But the children of these two great men were not to have the same kind of doughty optimism. Robert Darwin was one of Eras-mus's older children, the second to survive past the age of twenty-one. Though he was of squeamish bent, his father forced him to attend medical school at Edinburgh. And then, when it became nec-essary for Robert to buy into a medical practice, the assistance from his father was scant. This situation arose because Erasmus adored conceiving and raising children. Thus, on the death of Robert's mother, Erasmus set about fathering various bastards and then married again, keeping his illegitimate children with him, with additional children soon following. This was a considerable pack of mouths to feed, clothe, and house, and Erasmus had little time or funds for a grown-up son. Like a Californian caricature from the 1960s, Erasmus was profligate, charismatic, and randy, while somewhat lacking in foresight. Robert was left with little to receive by way of settlement or inheritance.

Robert had to make his way in the world unaided, eventually finding a country medical practice in Shrewsbury. His working life was characterized by a great aversion to the physical side of medical practice. He appears to have been an early psychotherapist, instead, talking his wealthy patients out of their misery. In addition, he very carefully husbanded and invested his funds, becoming an important capitalist and building a great family fortune. He lived quite abstemiously for his wealth, which was not fully comprehended, even by his own family, until some decades after Dr. Robert had become extremely rich. He was, in fact, something of a miser. But he loved his children deeply, despite a brusque manner. A tall man to begin with, he became immensely fat, crowding his environment with a quiet, saturnine bulk.

One of the major factors making for Robert Darwin's somewhat funereal attitude must have been his wife, Susannah or "Sukey." Susannah appears to have suffered from poor health her entire adult life, though quick-witted and extroverted. She spent very little time with her youngest son, Charles. Child-bearing was apparently very difficult for her, and might have occasioned post-partum depression. She died at the age of fifty-two, possibly of peritonitis, after some twenty-one years of marriage and six children. Robert Darwin never remarried.

THE YOUNG ANGLICAN NATURALIST

Periods of extravagance in Western Civilization are usually followed by periods of retrenchment and conservatism. When one contrasts the bacchanal of the 1960s and early 1970s with the subsequent conservatism of the 1980s and 1990s, one sees something of this contrast. But before the twentieth century, time flowed with greater leisure. The period from 1648 to 1789 could be described as a high point for European civilization. The subjugation and exploitation of much of the world was proceeding unchecked, from the wilds of the two American continents to the vast Siberian hinterlands of Russia. Plague was receding from the European experience, with the last major outbreaks coming in the early 1700s. Feudalism was dying, and everywhere prosperous townsfolk and landholders were building up a bumptious middle class. A flowering of thought and publication was continuing the momentum begun by the Renaissance. This was the gestation period for the modern world.[4]

Inevitably, the party came to an end, with a wooden thunk. The execution of the French royal family by guillotine in 1793 marked the end for many upper-class Europeans. Where before there had been a great continuum among the "progressives," as we might now call them, the beheading of Marie Antoinette soured the social climate. Whigs like Edmund Burke, who had defended the causes of American independence and Irish liberation, became staunch critics of revolution. William Pitt (Prime Minister, 1783–1801, 1804–6), otherwise a liberal, introduced emergency powers for the suppression of rebellion and suspended habeas corpus in 1794. Lord Liverpool himself watched the storming of the Bastille and the slaughter of the defending garrison during the French Revolution. The effect this had on his fifteen years (1812–27) as Prime Minister is clear; the public order was to be maintained at all costs. What before had been the rush of humanity toward ever greater enlightenment, peace, and freedom, in the eyes of Whigs and their allies, now threatened to become a pell-mell stampede in which all humane Christian values were in danger of being trampled, as far as the conservatives were concerned. It had become time to put an end to radical ideas and radical behavior, or at least to compromise them to the point of harmlessness.

Due to the madness of George III, the future George IV came to the monarchy via the compromised route of a Regency, in which the son ruled in place of the father. While the youthful sympathies of George IV had been with the Whigs, by mid-life he had given up any strong allegiance to them. Indeed, much of England was by that time disgusted with radical politics, particularly as it was associated with the excesses of the French Revolution and the loss of the American colonies. George IV satisfied himself with the pursuit of wine, women, and oriental curios. The last pursuit was physically embodied by the Royal Pavilion built for George in Brighton. In retrospect, the building can be seen as a forerunner of the taste and discretion of modern-day theme parks, done to George's taste.

The first thirty years of nineteenth-century England are known to us now overwhelmingly in terms of two great cultural landmarks. One is the stentorian history of the Napoleonic Wars. The other historical reference point is provided by the novels of Jane Austen, one of the greatest of English authors, despite her aversion to strenuous philosophizing and hysterical dramatization. In their place, Austen presents the orderly world of the Regency landed gentry. Unlike the period of the Enlightenment, the Regency was to be characterized by the avoidance of intellectual conflict and indeed ideas in general. This was the time when English charm and gentility were perfected, from dress to manners to euphemism. The elements had been there before, particularly in the Bath of Beau Brummel, but now they were to coalesce and lock English culture into a vise from which it would not escape until WWI. Rarely have so many people enjoyed such material comfort without the burden of despotically oppressing vassals. It was a time when an excess of well-educated, well-off but not actually rich, people could agonize about their amorous prospects. This era, and its fiction, has become one of the main influences on modern romance novels.

The Darwins and the Wedgwoods could have been lifted wholesale from their lives and plunked into the middle of one of Jane Austen's novels and no one would have been the wiser. In this we are singularly fortunate, because there are few feats of the imagination more difficult than understanding a person of a different historical period "from the inside." In particular, it is hard to understand a person before he or she has become "Great," and thus heavily documented by their own writings and those of others. This is the task that we are presented with when we try to understand the young Charles Darwin. This is where Jane Austen comes to our rescue.

Born in 1809, Charles Darwin had lost his mother by the age of eight. However, this unfortunate event seems to have had little effect on him. Her invalidism had left him hardly knowing her. He had been cared for instead by his older sisters and servants. By all reports, the young Charles was engaging, but not brilliant, affectionate, but not gushing. He had a broad face and a proper British beak of a nose, of which he was quite ashamed. He stammered and was somewhat spoiled by the sisters and servants. His father prevented him from going completely soft by sending the nine-year-old Darwin to a Dickensian boys school in Shrewsbury, in which privation, physical abuse, and classical learning were the main features of its traditional educational system. After that, Charles accompanied his older brother, Erasmus, to Edinburgh University in 1825, where they both studied medicine. This was the same medical school that two previous generations of Darwins had attended. Indeed, one of their uncles had died there of septicemia as a result of a classroom cut. Just like his father Robert, Charles found dissection and surgery nauseating, and had occasion to flee from the operating theaters. Charles was utterly repelled by medicine, and resolved to abandon the profession.

At this point, a very different element of Charles's background came to the fore. As a boy, he had had an intermittent passion for nature and science, much encouraged by both his father and older brother. At Edinburgh, this passion surfaced strongly. He began to take courses in natural history, and joined the Plinian Natural History Society, a student group, in 1826. It was at this time that Darwin started to exhibit his lifelong habit of forming friendships with professors and others interested in natural history. At that time he even made a minor contribution to a published work on marine biology, discovering some unusual specimens in the Firth of Forth. But Darwin's abandonment of medicine made his continuation at Edinburgh pointless, and he returned to the family home in 1827, at the age of eighteen.

At that point, Charles Darwin was a young Regency gentleman whose family had considerable wealth and comfort, but it remained unclear what profession was to be his in the world. His sisters were full of exhortations and his father, though taciturn, had high standards for his sons. Jane Austen affords us a snapshot of a fictional character from *Sense and Sensibility*,[5] Edward Ferras, which could be offered as a reasonable rendering of the young Charles Darwin:

> [He] was not recommended to their good opinion by any peculiar graces of person or address. He was not handsome, and his manners required intimacy to make them pleasing. He was too diffident to do justice to himself; but when his natural shyness was overcome, his behavior gave every indication of an open, affectionate heart. His

understanding was good, and his education had given it solid improvement.

It was at Cambridge University that Charles Darwin's life began to come together. He went up to university to become a clergyman, following his father's second choice for Charles's career, after medicine. At that time, in his own words, Charles "did not then in the least doubt the strict and literal truth of every word in the Bible."[6] Darwin was a lukewarm Anglican, and hardly as radical as his brother, father, or grandfather Erasmus, where religious matters were concerned.

Though Cambridge University around 1830 continued in the heavily classical vein of Darwin's boyhood education, he spent a great deal of his time and energy on natural history. His second cousin, William Darwin Fox (1805–80), introduced him to the joys of beetle-collecting, a promising hobby given the vast numbers of beetles to be collected. He also spent time hunting, drinking, and laying about, like a good upper-class twit. But in Darwin's case, these skills were to prove invaluable in his early career. On the other hand, Darwin also associated with his more scientifically oriented professors, particularly Botany Professor the Rev. J. S. Henslow (1796–1861), gathering knowledge in the course of conversation, and taking their courses.

Darwin's great opportunity to escape the tedium of English provincial life came in 1831. Captain Robert FitzRoy, a descendant of one of Charles II's illegitimate children, wanted a naturalist as a companion on his naval survey vessel, the *Beagle*, but it was important that the naturalist be a gentleman. Some lower-class drone who knew all the relevant biology and geology wasn't enough. No, it had to be someone from the gentry. Despite the fact that the main purpose of the voyage was to be the collection of natural history specimens and data, it was the class background of the ship's naturalist that was paramount. Here was where Darwin's background as a proper Regency layabout came in perfectly: he was the twit to fit FitzRoy's prig. Darwin badly wanted to go; his more charismatic professors had filled his head with dreams of travel to the tropics. His father rightly feared that this course of action would disrupt Charles from settling down to a quiet parsonage. But Charles's uncle, Josiah Wedgwood II, strongly favored the idea, and successfully persuaded Dr. Robert to let his son go.

The ship set out in December 1831, on a trip that was to last almost five years and circumnavigate the world. Darwin left a callow young man. He came back a proficient naturalist. He collected warehouses of biological specimens, many shot with his own hand or netted out of the ocean from the deck of the *Beagle*, together with rocks and fossils of all kinds. Most of the collected materials came from South America, that continent being the Royal Navy's particular interest for the trip. These materials constituted one of the more considerable treasures of natural history collected to that point.

But Darwin did not return to England in 1836 an evolutionist. He did not have any great insights into the origin of species while serving the Royal Navy. His main scientific speculation concerned the role of coral reefs in the formation of tropical islands, an idea that led to one of his first publications. This reflected the influence of Charles Lyell's *Principles of Geology*, which Darwin read on board the *Beagle*. The story that Darwin had the idea for evolution during the voyage of the *Beagle*, particularly on the Galapagos Islands, is a myth. Great ideas instead come while napping in suburban gardens or in studies.

A Theory by Which to Work

Upon his return to England in 1836, Darwin first settled in London. His exploits during the voyage of the *Beagle* had been spread about by family and academic friends. This gave him a new status. No longer a young man of uncertain parts, he had "arrived." He was made a Fellow of the Geological Society in 1836, and would later become its Secretary. In 1838, he was elected to the Athenaeum, the London highbrow's club. And in 1839, he was elected to the Royal Society, like his grandfather before him. In Charles's case, the chief basis for his scientific reputation was his extensive collections from the voyage of the *Beagle*. These were, at that time, one of the principal wonders of the London community of naturalists. Darwin began to circulate widely in intellectual circles, becoming acquainted with a range of notable figures, from Thomas Carlyle, a polemical historian, to Charles Babbage, the inventor of a mechanical computer.

In 1839, Darwin published his *Journal* describing the voyage of the *Beagle*.[7] This book was marked by fascinating natural history and lucid prose, becoming a best-seller for its day. On a more serious level, Darwin was busy editing the numerous monographs produced by specialists studying the specimens that he had collected on the voyage. It was this work which led him to the theory of evolution. The crucial finding was quite inadvertent. Darwin and others on the *Beagle* had collected numerous bird specimens while in the Galapagos, including mockingbirds and finches, noting the island from which each came. Darwin had thought that many of the different forms were mere varieties, rather than species. But zoologist John Gould, who knew much more about birds than Darwin did, concluded in March 1837 that different islands had different species of birds. Interestingly, all of these birds seemed to be similar to mainland forms from South America, from which the ancestors of these birds could have flown to the Galapagos Islands. This was a bombshell for Darwin, who had been uncertain as to the status of the birds. He quickly deflected his uncertainty onto the phenomena themselves, concluding that the island species had somehow been derived from the birds of the mainland species that had come to the Galapagos Islands, by a process of gradual "transmutation." Soon he hit upon the idea of a "tree" connecting different life-forms, each undergoing transmutation or, as we would say, evolution.

Darwin was by that time a sophisticated man of science. He knew that it wouldn't do to merely suppose some unspecified transmutation process. It was also incumbent upon him to discover a well-defined mechanism by which evolution could proceed: "a theory to work with." Solving this problem racked his brain, and he spent feverish months trying out one idea after another. The next breakthrough came in September 1838, when Darwin read Thomas Robert Malthus's work on population growth.[8] The economist Malthus was concerned that human population size tended to grow geometrically, while food supplies increased at most in a linear fashion, or so he assumed. But granting this assumption, it is then inevitable that population size would come to exceed the supply of food, resulting in famine, disease, and generally disagreeable behavior (war, cannibalism, etc.). In so reasoning, Malthus would make his reputation as one of the founders of economics as a "dismal science." Darwin was interested in the consequences such a calamity would have for different members of populations of plants and animals. His reasoning was that the better-adapted organisms would survive this calamity, and thus be better represented among the parents of the next generation. Granting only the supposition that like gives rise to like, the offspring of these superior survivors should themselves be superior, at least by a small increment. Carried on generation after generation, this process would lead to natural selection of fitter descendants.

The Great Procrastinator: Marriage and the Origin

The problem with keeping detailed notebooks of one's thoughts is that they can later prove to be very embarrassing. How would Darwin feel if he knew how many biographers have satirized his handwritten agonies about marriage? According to his notebooks, he didn't want to be a neutered worker bee, with nothing but work. But marriage threatened to involve children, quarreling, and, worst of all, less time for work. And then there was the risk that his wife wouldn't like living in the city, and would banish him to the idiocy of rural life, a sentiment Darwin then shared with Karl Marx. On the other hand, marriage would supply someone to take care of the house, "chit-chat," music, and so on.

But there were other issues as well. Darwin's success as a scientific personage in London had greatly impressed his father, so there was no more talk of Darwin becoming a country parson. Dr. Robert Darwin was willing to support Darwin and his scientific work even if he married, removing the obstacle of parental disapproval. As a married man, Darwin would be allowed to remain without profession.

Perhaps a point of greater importance was whom was Charles to marry? He had spent almost his entire adulthood in the company of men, most especially on board the *Beagle* for five years. Even in London, he had not frequented fashionable parties or balls. Being a member of the landed gentry, however, tended to solve problems like these. Families, family connections, and family friends were everything. In particular, he had been infatuated with the Wedgwood family and its convivial socializing most of his life, not only visiting with them, but also corresponding frequently. His mind made up, his father's permission secured, in 1838 he set out for the Wedgwood home, Maer, and began to court Emma Wedgwood. The marriage took place in January 1839.

The young Emma was attractive and lively, if her portraits and contemporaries are to be believed. But she was also Charles Darwin's first cousin on his mother's side, just a few months older than Charles, the youngest daughter of Josiah Wedgwood II. "Josiah II" was the man who got Darwin on board the *Beagle*, over Dr. Robert Darwin's objections. Charles's sister Caroline had married Emma's brother Josiah Wedgwood III. Genetically, this is all rather unsettling, since first cousin marriages give progeny with a greatly increased frequency of birth defects, among other problems. But the Anglican rite does not frown on the practice, and the nineteenth-century English landed gentry were nothing if not Anglican.

Around the time of his arrival in London, Charles began to suffer from a variety of baffling medical complaints. These included heart palpitations, gastric upsets, and headaches. Almost all of 1840 was given over to ill-health. This year apparently persuaded the Darwins that they had to move to the country for the sake of Charles's health. So, despite his earlier fears of being dragged off to the countryside by a relentless wife, in 1842 Charles shuffled off to Down House, in Kent, quite willingly. Notwithstanding the medical motivation, in clear country air Darwin suffered ill-health for the rest of his life. However, in the Victorian tradition, he suffered with conspicuous bravery, and a little morbid obsessiveness, until the age of seventy-three, producing thousands of pages of scientific manuscripts en route to his much-delayed grave.

Some have attributed Darwin's ill-health to neurotic hypochondria arising from fear about the reception that his theory of evolution would receive. Darwin knew that his views on the transmutation of species, or evolution, flew in the face of settled opinion among both clergy and the scientific laity. And being a stammering, sincere kind of Regency gentleman, the last thing that he was suited to was causing a social rupture. This was a man who could deal with his father only on terms of virtual prostration. That Darwin would repeatedly delay the publication of his theory of evolution by natural selection is not the least surprising. Grandfather Erasmus Darwin, on the other hand, would have published and be damned.

However, Charles Darwin was also concerned that his great achievement should be acknowledged by future generations. With that in mind, he prepared two brief sketches of his ideas, one in 1842, and again at greater length in 1844. The second work was about fifty thousand words. He arranged for its publication in the event of his death. While this work was never published, he did show it to some scientific friends, though by no means all of his scientific acquaintances. Many of them did not learn of Darwin's theories until they read the first edition of the *Origin of Species*, in 1859.

In the years between 1844 and 1855, Darwin set about developing his understanding of biological detail, most notably in a mammoth study of barnacles which resulted in the publication of two thick volumes on this most adhesive of crustaceans. During this time he received the Royal Medal from the Royal Society for his work on coral reefs and barnacles. By 1855–56, he was finally settling down to write an even more massive presentation of his ideas about evolution. Though his colleagues tried to persuade him to publish a brief summary first, he was adamant about the need to be truly exhaustive. As events turned out, Darwin was not going to have any leisure for further hesitation.

And then it finally happened. In June of 1858 Darwin received a letter from zoologist Alfred Russel Wallace, together with a brief paper outlining a theory of variation and natural selection together acting to produce evolution. Wallace was a collector of natural history specimens who traveled through the tropics, as well as an author of scientific papers and a fairly popular book somewhat in the same spirit as Darwin's *Beagle* journal. (William Adamson, the hero of A. S. Byatt's novella *Morpho Eugenia* [later made into the film *Angels and Insects*] shares some biographical features with Wallace, including a relatively modest background and the loss of a ship's worth of specimens at sea.)[9]

Darwin had a great crisis when he read Wallace's manuscript, because he had been asked by Wallace to communicate it further to Sir Charles Lyell, with a view to publication if deemed worthy. But Darwin didn't want Wallace to be known as the sole discoverer of the idea that he, Darwin, had been sitting on for almost twenty years. Darwin appealed to Lyell for advice, and Lyell suggested that Wallace's paper be communicated together with one from Darwin, so that they could have codiscoverer status. Accordingly, two papers were presented before the Linnean Society in 1858, one from Wallace and one from Darwin. Neither author was present, and there was little discussion.

Wallace having burst his bubble, and being left in public view on the evolutionary question, Darwin hurried to come up with something that would defend his views better than a mere paper could. Thus, he prepared an "abstract" of a few hundred pages, the *Origin of Species*, first published in 1859. Subsequent generations must be grateful to Wallace, because Darwin's original plan would have run to many volumes. While Darwin was a fairly good writer by Victorian standards, it would have been unlikely that any but the most dedicated would have bothered to read his presentation of evolution in the form he originally intended. As it turned out, the *Origin* is a historical milestone: a book that revolutionized a large scientific field, but which intelligent lay people can read profitably.

The *Origin* is notable for the extent to which it directly conveys the structure of its own arguments and likewise the extent to which presumed opponents of these arguments are dispatched by substantive refutations. It does not dutifully and painfully muster particular "facts" of natural history, one point at a time. Rather, it has a boldness and clarity that are almost shocking compared to other scientific publications of its day.

The reaction to the *Origin* is famous in the history of science. Some pompous prelates attacked it, while rising young scientific stars rallied in its defense. T. H. Huxley made a reputation for himself as "Darwin's Bulldog" by defending Darwin against Bishop Samuel Wilberforce ("Soapy Sam"). Magazines published caricatures of Darwin. Darwin became a cult hero for leftists, revolutionaries, anticlericals, and the like. Karl Marx wanted to dedicate *Das Kapital* to Darwin, who politely declined. Darwin certainly had no idea of the bizarre historical junction that this proposal constituted, an exchange between the two most important thinkers of the nineteenth century. The scientific community rather liked the idea of evolution, but the concept of natural selection proved more difficult to swallow. However, such things are the concern of the next three chapters.

On to WESTMINSTER ABBEY AND IMMORTALITY

What could Darwin do for an encore? As discussed in chapter 2, he went on to wrestle extensively with the problem of heritable variation. Though he probably was as expert in that field as anyone then alive, bar one—Mendel—he never properly sorted out the mechanics of inheritance.

Darwin also worked on other biological problems that were of interest to him, such as sexual selection, orchid fertilization, animal behavior, earthworms, and the evolution of humans. Darwin's views on man were somewhat parochial, but he had some impressive intuitions, such as the location of human origins in Africa. All his work was characterized by superb attention to both detail and significance.

Though to contemporary eyes having ten children, the last when Emma was forty-eight, seems like an ungodly number, such fecundity was in fact fairly common in the era before birth control and modern hygiene. And as in that era generally, it is not surprising that two children died in infancy and one in childhood. Whether the genetic problems of a first-cousin marriage played a role in the medical problems of their children is to some extent unknowable. Anecdote has it that the Darwin family had a number of odd but persistent medical complaints. Some authors are of the opinion that the Darwins were a pack of hypochondriacs.

Of perhaps greater psychological significance was the long-standing issue of religion within the Darwin family. Charles Darwin gradually became an atheist, while his wife remained devout her entire life. Darwin died believing that that would be the end of him, for good and all. There are stories of a deathbed conversion to Christianity by Darwin, but they are entirely apocryphal. Darwin was buried in Westminster Abbey, his corpse finding a final resting place near the bones of Isaac Newton, one of his few peers in history. His gravestone gives his name, date of birth, 12 February 1809, and date of death, 19 April 1882. While others entombed around him have been accorded panegyrics chiseled into stone, his stone has none.

What then is Darwin's historical standing as a scientist? There are three broad features of life which were puzzling before Darwin: relatedness of species, diversity of species, and adaptedness of species. Darwin provided the basic explanations for these phenomena still used by biologists today. Each of these points will be central to this book, but their essential outline is easy to provide now.

The problem of the similarity of mammalian species to each other, or insect species to other insect species, was one which, before Darwin, was solved only by appeal to theological or philosophical ideas. Darwin's solution to this problem is that species are related evolutionarily, so that they resemble each other in something like the way three brothers resemble each other; they have a common ancestry. On the long-term evolutionary scale, species descend from each other by a process of slow and gradual change, in which the relics of common ancestry will usually be visible. Ultimately, all life is to be traced back to one or a few original ancestors, whose features define some of the basic limits for life itself.

Life is perversely diverse. Why should there be over half a million species of beetles, far more than the number of terrestrial vertebrate species? Why should life be so abundant, so varied, so lavish? And then there is the problem that these different species are not necessarily just minor variations on a theme, though such of course exist. Some mammals fly, some swim in the deep ocean, while others burrow in the ground. This too Darwin was able to explain using the same branching tree of evolutionary derivation. From the branching comes diversity, and often selection will foster this diversity, pushing species to great extremes to escape ecological competition in the ancestral habitat. (This will be explained more later.) In Darwin's scheme, diversity becomes not merely explicable, but unsurprising.

Finally, there is the adaptedness of life, the way it seems to operate by marvelously efficient contrivances. This was the seat of the strongest argument for the existence of God known to the pre-Darwinian mind, the argument from design. This argument amounts to the necessity of invoking a creator to explain these marvelous contrivances, given the implausibility of their arising by accident. This adaptedness Darwin explained using natural selection, the differential reproduction of those with varying fittedness to the environment. The faster horse that can outrun its predators will tend to have more progeny, on average, and progeny somewhat like itself. Thus the average speed of the horses increase. Darwin's natural selection made adaptedness such a plausible result of evolution that some biologists have if anything a tendency to overdo it, seeing adaptation everywhere. Whatever the errors that such scientists fall into, adaptation is in no way a difficult problem for biologists to explain using evolution by natural selection.

In total, Darwin provided a foundation for biology which was completely free of religious elements. For molecular biologists, this contribution is very much in the background. They can just go on with experiments whose rationale derives more from principles of organic chemistry than evolutionary biology. But the fact that they do not need to bow toward Rome or Canterbury, by way of scientific piety, is largely the result of Darwin's achievement. And away from those biological fields where straightforward molecular analysis is enough to unravel all problems, the evolutionary reasoning first provided by Charles Darwin is often the central means of intellectual analysis, defining problems, suggesting alternative solutions, and deciding among competing hypotheses. Modern biology would be inconceivable without Darwinian theories and findings.

2 HEREDITY

The Problem of Variation

THE HARDEST THING to understand about biology is the importance of variation. There are probably several reasons for this. One may be human psychology. It is easier for us to think in terms of homogeneous classes of things, rather than heterogeneous sets of things. If we count out three plastic yellow ducks for our toddler, then we are implicitly conveying the idea that they are the same thing, which is of course illusory. The yellow ducks are bound to be different from each other, at least in subtle ways.

Another problem is that the foundations of science were defined by physicists using mathematical representations of reality in which any type of variation or randomness was neglected. This is the metaphysics by which integral and differential calculus become cosmology. Elegance, precision, and generality were brought together by Galileo, Newton, and Laplace to an extent that had never been seen before. And their highly abstract theories became the model, the paradigm, for the construction of science.

But it turns out that these important tendencies of human thought, and the thought of moderns particularly, are highly counterproductive in biology. Variation is not merely characteristic of living things, it is also essential to their very evolution. Darwin was the first person to see this clearly. Unfortunately, Darwin's understanding of heredity was not enough for him to work out the specific mechanisms by which variation contributes to evolution. He got the first part, the importance of heritable variation. But neither he nor any other prominent biologist in his lifetime worked out the mechanics of inheritance. This led him into many erroneous inferences about the workings of evolution.

The key to the mystery of heredity was found by a nineteenth-century Austrian monk, Gregor Mendel, who worked in virtual obscurity. Placing that key into the lock of variation, and opening the door of evolution, would not be accomplished until after WWI. But after many plot complications, it was finally understood that the actual mechanism of heredity makes Darwinian evolution possible.

The Platonic Organism

Biology was not founded with a reasonable understanding of variation. Instead, the academic roots of biology come directly from a source entirely inimical to a proper appreciation for variation, the philosophy of Plato, the founder of academia and academic knowledge.

It is almost inherent to academic knowledge that it abstracts, neglecting distracting variation. The real and substantial is represented by words or symbols. To some extent this property is in herent in language itself. But in academic knowledge it is carried much farther, particularly in the use of abstracting concepts. Not only do we have words for the children and animals that we know, we also have words which refer to general categories like child, horse, and, for that matter, word. And in academic discourse, such abstract words are piled on top of one another in towering heaps of verbosity. You can practically define an intellectual as someone who is so infatuated with the general that the specific is lost to awareness.

The great epistemological problem is how abstractions or generalizations can improve knowledge. Concrete statements seem obviously more realistic. We can say that John Kramer crossed the street in front of his house on Tuesday, and clearly, when this is true, we have said something of definite informational value. Compare this with statements like "man is a mammal." Where is the value in statements of this kind?

This was a problem which greatly vexed classical Greek philosophers, including Plato. Alfred North Whitehead, among others, referred to Western philosophy as consisting primarily of a series of footnotes to Plato. And Plato had a solution to this problem of abstract knowledge. His theory was that underlying the surface of things are their true inner natures. Such natures are made up of fixed "ideas," or attributes, which are only obscured by the variety of ways in which individuals manifest them. Thus, different olive trees are only variations on the essential or true olive tree. Abstract knowledge is "truer" for Plato because the ephemeral, trivial, and distracting superficial attributes are ignored, leaving only the pure idea.

In biology, Plato's idea of an underlying essence has had overwhelming appeal, particularly as presented by the first great biologist, Plato's student Aristotle. While any species may possess numerous individuals, they all seem to conform to some underlying plan of morphology and behavior. Classical biology assumed that there is an underlying species "essence," with individual variations of no meaning, "accidental deviations from the type." The essence, or "eidos," was seen by Aristotle as transmitted from parent to offspring. This was one of the original sources for the concept of inheritance.

This conception also fit well with the theology of Western Christendom after the Renaissance. If God is conceived as an all-powerful being capable of the most perfect creations, then the species that He creates should reflect His original intentions in their underlying nature. There might admittedly be accidental deviations from the initial divine conception, but careful study of multiple specimens should reveal the originating impulse in all its coherence and function. Essences and Supreme Beings go together quite nicely, lending an elevated tone to biology, which after the Renaissance fashioned itself part of "natural theology."

When the true representatives of the species are thought to have certain ideal attributes, and deviations from them are accidental or "impure" variations, then evolution is unlikely to be thought of or accepted. Variation becomes distracting noise, not material for change. And since species are thought of as a set of discrete types, the notion of hopping from one species to the other becomes the only conceivable form of change. Such hopping in turn is quite implausible, and indeed seemingly miraculous. This made the development of evolutionary reasoning, even by those who were open-minded, extremely difficult. Equally important, it also made those who could not reason evolutionarily still more closed-minded in their opposition to evolution.

Darwin's Emphasis on Variation

Collecting specimens and fossils for display in one's study was an abiding passion among Victorian ladies and gentlemen. Recall that Charles Darwin was an avid beetle collector at university, even as an indolent undergraduate. He was fond of telling the story of how, one day, he had caught two different beetles and was holding one live beetle in each hand when he saw a new beetle which he just had to have. Desperate, he popped one beetle into his mouth, and reached for the third. Before his fingers could close on the new specimen, the beetle in his mouth ejected an acrid, stinging fluid. Darwin spat out the beetle, losing the one remaining in his hand too. End result: no beetles.

In any case, Darwin had a lot of beetles even before he set off on the voyage of the *Beagle*. Then on the *Beagle* he was to collect numerous specimens of just about everything he could find, from birds to plants to marine life. He also spent about ten years, from the mid-1840s to the mid-1850s working on the taxonomy of barnacles. Darwin corresponded with animal breeders, especially pigeon fanciers. All of this work exposed him to the fantastic diversity to be found within a given species. Indeed, in preparing his species classifications, he often found it exasperatingly hard to sort out what was what, because the life-forms were so variable.

Variation was the starting point for Darwin. In the *Origin*, the first two chapters are devoted to variation, one under domestication, the other in nature. For Darwin, variation was obviously important:[1]

> . . . it is really surprising to note the endless points in structure and constitution in which the varieties and subvarieties differ slightly from each other. The whole organization seems to have become plastic, and tends to depart in some small degree from that of the parental type.

It is interesting that Darwin expresses surprise at this situation, perhaps reflecting its divergence from Platonic expectations. The allusion to "parental type" suggests a lingering residue of typological thinking. In any case, the assertion of the importance of variation is clear.

Just as important for Darwin's argument is the inheritance of variation.

> Any variation which is not inherited is unimportant for us. But the number and diversity of inheritable deviations of structure, both those of slight and those of considerable physiological importance, is endless.

The animal that Darwin discussed in greatest detail in the *Origin*, where variation was concerned, was the domestic pigeon. Indeed, pigeon breeding was a major hobby among the Victorian working classes. Darwin made it his own too, and became a great expert in pigeon breeding. A reviewer of the *Origin* before publication was unimpressed with the evolutionary arguments but quite liked the pigeon material. His advice to the publisher was, instead of publishing the *Origin*, to request a book on pigeons from Darwin. In any case, Darwin used the many varieties of pigeon to argue for the existence of variation amenable to selective breeding for special attributes, and thus diversification, in effect a kind of model for the evolutionary process as a whole.

Later Darwin was to prepare an enormous treatise on variation, particularly with regard to patterns of inheritance. This treatise was *The Variation of Animals and Plants under Domestication*, appearing in 1868. Unfortunately, Darwin's attempts to work out the pattern of inheritance failed. Sadly, an unopened copy of Mendel's crucial paper on inheritance in peas was found among Darwin's files after his death. If he had read it with understanding, evolutionary biology would have gained at least three decades.

The main alternative to any theory of evolution is a static living world, which presumes that all species cannot change. This then naturally requires that species be created *de novo*, since they cannot change into one another. Since a reasonable inference is that living things couldn't accidentally come together out of the blue, by purely mechanical accident, it is more plausible that species were created by an omnipotent intelligence, if stasis prevails. Thus creationism and stasis are natural corollaries of one another, while materialism and evolution are similarly naturally allied. Variation is the hinge on which biology turned from creation and stasis to adaptation and evolution.

DARWIN NEEDED MENDEL

The best way to convey the importance of Mendel's genetics as a mechanism of heredity is to consider some of the alternative ways in which variation could be transmitted from generation to generation. In other words, to consider alternative mechanisms for heredity. This is in fact more than an imaginative exercise. Some of the alternatives to be presented here had strong adherents through the latter half of the nineteenth century and well on into the twentieth century. It is a major historical irony that one of the most articulate proponents of these erroneous views in the nineteenth century was none other than Charles Darwin himself.

The first of Darwin's pratfalls was the theory of "blending inheritance." The term blending inheritance originally had no clear meaning. In a sense, this theory is the more abstract version of the general notion of "blood" current in Europe before 1900. Thus, there was "royal blood," "aristocratic blood," and so on. This concept of blood remains as a folk theory of inheritance among those without a biological education, especially in the sense of "bad blood," a loose concept that can refer to hereditary disease, ill-feeling, or original sin.

The scientific concept of blending inheritance is less dramatic. We'll begin with a hypothetical example. If we imagine a white rabbit crossed with a black rabbit, on the blending theory of inheritance all the offspring would be gray. If the gray rabbits were crossed with each other, all their offspring would also be gray. In fact, so long as you crossed like with like, there should never be any chance of getting anything different. In this way, characters breed true.

Blending inheritance is basically inimical to evolutionary change. The reason for this is that it destroys variation. This point is easily seen if we return to our rabbits. With blending inheritance, when white rabbits mate with white, we get more white rabbits. Similarly for black mated to black. When black mates with white, we get gray rabbits. This might seem like more variation than we started with, but this type of outcome means that fewer rabbits are at the extremes of white or black. Furthermore, gray crossed with white will produce gray, and never white, while gray crossed with black will produce gray, and never black. Through time, over generations of mating, if the pairing of rabbits is roughly random with respect to color, there will be fewer and fewer white or black rabbits. Everything will become a shade of gray. And then, as these grays mate with each other over more generations, their fur color will converge to a uniform gray, with no variation remaining in the population. The extremes will be lost, leaving the population with nothing but a dull uniformity. Once every individual is the same shade of gray, there is no variability to contribute to evolution, and stasis is the likely result. Thus blending inheritance is unfavorable for Darwin's theory of evolution.

This problem was pointed out to Darwin by Fleeming Jenkin, a British engineer.[2] Darwin regarded it as an extremely serious difficulty, one for which he never really developed a solution. His strategy was to look for ways of generating an abundant supply of new variants, on which natural selection could act. This supply was to come from the inheritance of acquired characters.

The inheritance of acquired characters is the idea that what is transmitted to offspring is dependent on the environment or on the physiological state of the parent. That is, the attributes inherited by offspring are influenced by corresponding attributes acquired by their parents during their lifetime, as a result of environment or actions. For example, if there is inheritance of acquired characters, then the children of parents with numerous tattoos should be born with markings on their skin. Two particular types of such "soft" inheritance were commonly discussed before the twentieth century. First, it was thought that if a parent has "striven" to achieve a particular goal, such as a giraffe stretching to reach leaves atop a tree, then it would develop appropriate advantageous structures during its lifetime that would then be transmitted to its progeny. Thus, the stretching parent giraffes have baby giraffes with longer necks. Second, it was thought that the circumstances of reproduction had a strong influence on the nature of progeny. For example, Francis Bacon in the early part of the seventeenth century thought that the circumstances of conception influenced the lifespan of offspring, especially when fathers are[3]

> full or drunk; others after sleep, or in the morning; some again after a long intermission, and others after a frequent repetition of the conjugal act; some (as generally in the case of bastards) in the heat of passion.

Indeed, both of these sorts of belief are common to all cultures which have not had genetics take over from "common sense" in discussions of heredity.

In short, the idea is that inherited material, "the germ plasm," is sensitive or ductile. Supposedly, the germ plasm takes on impressions from its surroundings, and in so doing shapes the next generation. In particular, it was commonly thought that the response of the next generation would be beneficial. That is, the next generation would possess enhanced powers of survival, competition, or reproduction as a result of the inheritance of acquired characters. There would thus be a kind of "adaptation" to the environment. Jean-

Baptiste de Lamarck (1744–1829), a leading French biologist, made this type of adaptation central to his conception of species transformation in nature. (Some consider him a forerunner of Darwin with respect to evolution.) With such a powerful mechanism of adaptation, Lamarck could propose a theory of adaptive evolution which lacked any idea of selection, some form of vital striving taking the place of Darwin's natural selection.

Darwin's beliefs were different from Lamarck's. In particular, he did not conflate the process of adaptation with the process of inheritance. Instead, he used the inheritance of acquired characters as a source of new variation, variation that was not biased in the direction of adaptation. (His theory was that natural selection would then sort this variation to produce adaptation. This will be discussed in more detail in chapter 3.) With respect to the overall mechanism of heredity, Darwin developed an elaborate theory which he called "pangenesis." It combined blending inheritance with the inheritance of acquired characters. He supposed that the material of inheritance was made up of a large number of "gemmules," which were the particles of heredity. However, these hypothetical gemmules were not fixed in physical structures, and there could be a variable number of them of any one type. Darwin thought that gemmules migrated through the body, picking up information about the state of different body parts. Before reproduction, some gemmules would migrate back to the gonads, where they would then get on board the gamete prior to fertilization.

There are major problems with the theory that acquired characters are normally inherited. As we will see, Darwin's own attempts to test the theory led to its refutation. More important, contemporary research in genetics is utterly destructive of any theory based on the inheritance of acquired characters. It is true that irradiation or mutagenic chemicals affect the progeny of the next generation, but they do so in an essentially random way. There is no "impression of the environment" with mutation. There is only chemical disruption of an undirected kind.

From Gradualism to Galton

One of the basic articles of faith in Darwin's training as a scientist was gradualism. Gradualism was a compound of two logically independent views of the world. The first, simply enough, was that processes of change are gradual, with no abrupt transitions. Thus, Darwin explained coral atolls in terms of the gradual subsidence of coral reefs around volcanic islands. In the geology of Darwin's day, subsidence, erosion, and sedimentation were held to be the key processes determining land-forms, each of them slow processes of gradually accumulating changes.

The second component of gradualism was a disbelief in outside intervention. Gradual terrestrial processes were appealed to as explanations for change, not cataclysmic changes coming from the heavens, whether astronomical or religious. No God, no astrological forces, and no cosmic spirit was to be used in causal explanation. These two components show why the gradualism that Darwin learned as a young naturalist led naturally to his theory of evolution. A Divine Creator of all animate forms is exactly what a gradualist would not like in an explanation of life. Thus, gradualism led Darwin to overthrow the then-conventional, theologically founded, biological theory.

The problem that arose from this twofold nature of gradualism was that anything that seemed to involve discrete jumps in inheritance was associated with a belief in the action of gods or other occult forces. This association was not legitimate. It arose from the nature of gradualism as a kind of scientific prejudice. Darwin could not, therefore, accept any type of inheritance except that which involved blending and continuous inheritance. The cases of bizarre mutant forms that were known in his time, he dismissed as "sports," which could not make any further contribution to evolution. In this he was largely right, since aberrant mutants normally have low viability. But the sports did indicate that inheritance wasn't necessarily what Darwin supposed. Darwin, like most scientists, sometimes took observations which threatened his theories as enemies to be combated, rather than as revealing exceptions. He himself confessed that he was a great "wiggler," meaning that he was good at wiggling off hooks threatening his ideas, that seemed to catch him in error. Indeed, most of the hooks upon which Darwin might have been impaled have proven to be spurious after further research. But not the hook of inheritance.

Darwinian orthodoxy was that inheritance was continuous. A small subfield grew up, "biometrics," devoted to the quantitative study of inheritance, presuming that it was continuous. The first great figure in biometrics was Sir Francis Galton, who is known primarily for his *Hereditary Genius* (1869).[4] In this and other works, Galton began to study variation statistically, creating the foundations for statistics, particularly as applied to the problem of inheritance.

In addition, Galton did some experimental work on inheritance. He performed blood transfusion experiments with rabbits, in which he showed that there was no evidence for the circulating gemmules of Darwin's pangenesis. The blood of rabbits of one color fur did not change the color of the fur of the offspring of rabbits of another color of fur that received that blood by transfusion. While Darwin encouraged Galton in these experiments initially, once the negative results came in Darwin argued that the gemmules could be circulated in some bodily fluid other than the blood. (A classic piece of wiggling.) Thus, while Galton developed some of the first tools for analyzing quantitative variation statistically, he came to believe that Darwin's view of heredity was mostly mistaken.

In late Victorian England, Galton was the leading figure in the study of variation from the standpoint of evolution. A variety of younger men attached themselves to him, most notably Karl Pearson and William Bateson. Pearson was primarily a mathematician by training, while Bateson had difficulty with mathematics. Pearson was attracted to Galton's statistics, and developed it further. Bateson was more interested in the idea of sports, and the prospect of evolution by "jerks" or saltations. (Saltation refers to leaping, from the Latin *saltus*, a leap.) As Pearson and Bateson were both ambitious and abrasive, they soon began to hate each other intensely. The stage was set for one of the most destructive episodes in the history of biology.

REDISCOVERY OF MENDELISM

The lit fuse was supplied by the rediscovery of the work of the humble Mendel. Mendel did his superb work on inheritance in peas from 1856 to 1871, publishing first in 1865, in German, in the journal of a local natural history society in central Europe.[5] One view of his undeserved obscurity would be that Mendel simply wasn't noticed. The truth is much worse. Mendel actively corresponded with one of the leading botanists of his day, C. Nägeli, who was a firm believer in blending inheritance. Nägeli appears to have deliberately discouraged Mendel, misled him about the results of other relevant scientific work, and seen to it that few other scientists heard of him. In so doing, Nägeli achieved almost complete suppression of Mendel's demonstration of discontinuous inheritance for more than thirty years.

It might also be noticed that this episode reveals the extent to which self-promotion is crucial to scientific success. Close study of the lives of Isaac Newton, Charles Darwin, or James Watson indicates that science is hardly the objective enterprise that its mythology claims it is. Those three scientists worked very hard to ensure that they were appreciated. (Only James Watson was honest about this.)[6] In fact, one might argue that self-effacement, which was required of monks like Mendel, is counterproductive in science.

But by a remarkable coincidence, a number of European scientists independently rediscovered Mendel around 1900. The story that his wife told after his death was that on May 8, 1900, William Bateson was riding on a train on his way to give a talk to the Royal Horticultural Society of England on "Problems of Heredity." He had brought along some papers to read, among them Mendel's. He realized soon enough that Mendel's paper was critical. The basic ideas were immediately incorporated into the talk that he gave that day. More recently, historians of science have argued that Bateson had already read Mendel's paper, but Mrs. Bateson's account is too quaint to avoid repeating.

To understand the lightning bolt that was Mendel's rediscovery, some of the basic elements of genetics have to be reviewed. Mendel proposed a model for heredity that was at once devastatingly simple, but also extremely good at explaining many of the puzzling results of plant and animal breeding experiments. Mendel's model was based on the assumption that inheritance is divided up into particulate units, now called "genes," each of which determines a particular character. Thus, we have the genetic slang that there is "a gene for eye-color," "a gene for dwarfism," and so on. The interesting point is that most organisms have two copies, or "alleles," of each of these genes. The concept of an allele refers to the specific state or type of the gene in question. Thus, you may have two cars, and these cars are your cars in the same sense that your genes are your genes. But the cars have their own particular make and model, which would define them as "alleles," in terms of this analogy. So saying that you have cars is to say that you have genes. To say that you have a Ford and a Chevrolet is to describe their allelic state.

The central point in genetics is that the two alleles of each gene (the two cars) are transmitted to offspring with equal frequency. This process doesn't matter much when these two alleles are the same, say they both code for blue eyes, but it does matter when they code for different characters, say one for blue eyes, and the other for brown eyes. (Or when one car is a Ford and one is a Chevrolet.) Furthermore, Mendel knew that combinations of alleles did not blend with or contaminate each other; they remained distinct. (Your cars have no tendency to become like each other.) For this reason, Mendelian heredity is called particulate and "hard," the exact opposite of Darwin's scheme of blending inheritance and the inheritance of acquired characters.

Further complications arise when different alleles are combined together at a particular genetic site, called a "locus." Continuing with our analogy, the locus is your two-car garage, with the two alleles at a locus being two cars inside the garage. "Sex" occurs when the cars leave the garage and join together in new combinations with other cars at new two-car garages. But when two different alleles have to express the character that their gene codes for, there are multiple possible outcomes. One is that the character becomes an *average* of what it would be compared to the characters generated by the two pure combinations of allele types. So if one allele makes a dog weigh sixty pounds when there are two copies present, and another allele makes a dog weigh eighty pounds when it is present in two copies, the combination of the two different alleles would be a seventy-pound dog. But there is another possibility, called *dominance*. This would arise if the two different alleles in this example, when combined, produce a dog of eighty pounds. The "heavy-dog" allele is then said to be dominant over the "light-dog" allele. Conversely, the "light-dog" allele is said to be *recessive* relative to the "heavy-dog" allele. Many characters exhibit this kind of genetic dominance, among them most alleles for pigmentation in mammals, whether eye color or fur color. In plants, like the peas that Mendel bred, this type of dominance is so common among characters, that Mendel thought it was a law of inheritance. But it isn't. Most organismal characters, like height or weight, exhibit intermediate values when two different alleles are combined.

William Bateson, along with many others, concluded that Mendel's model of hard, particulate, discontinuous variation defined the type of variation which was important in evolution. He also came to the conclusion that this showed that evolution itself must proceed in a saltatory fashion. Karl Pearson and the other biometricians who studied inheritance based on the concept of blending inheritance accepted this idea of an opposition between Mendelism and Darwinism. Their faith in Darwin's hypothesis of evolution by natural selection acting on continuous variation was absolute. To reject any part of it, for the biometricians, was to reject the whole. So they accepted Bateson's connection of the material of evolution with the pattern of evolution, the association of continuous variation with gradual evolution. Since Mendel's theory was one of discrete variation, biometricians concluded that Mendelism must be false. And therefore they attacked Mendelism.

Bateson was trenchant on the counterattack, accusing his opponents of "perverse inference," "slovenly argument," and "misuse of authorities, reiterated and grotesque." The biometric journals began to refuse to publish any of his articles. Matters came to a head at a meeting of the Zoology Section of the British Association for the Advancement of Science. Near the end, Karl Pearson rose to propose a truce:[7]

> On Pearson resuming his seat, the Chairman, the Rev. T. R. Stebbing, a mild and benevolent looking little figure for a great carcinologist, rose to conclude the discussion. In a preamble he deplored the feelings that had been aroused, and assured us that as a man of peace such controversy was little to his taste. We all began fidgeting at what promised to be a tame conclusion to so spirited a meeting, especially when he came to deal with Pearson's suggestion of a truce. . . . 'You have all heard,' said he, 'what Professor Pearson has suggested' [pause], and then with a sudden rise of voice, 'But what I say is let them fight it out.'

MUTATIONISM

The biometricians could not hope to beat the Mendelians where heredity was concerned. As a factual matter, the biometricians were wrong about the machinery of inheritance. It was particulate and "hard," as Mendel had proposed. Thus, the Mendelians soon found themselves triumphant outside of a small redoubt of biometrics. The problem that remained for the Mendelians was to develop their own theory of evolution, since they could have no part of the selection-oriented reasoning of the biometricians, given the spurious association of Darwin's theory of adaptation with his theory of blending inheritance, a model of inheritance that is in fact inimical to evolution by natural selection.

The key figure in this further development of Mendelism was Hugo DeVries, a botanist who experimented with the evening primrose, *Oenothera lamarckiana*. (Just the name of the plant is enough to send a chill through a Darwinian.) This plant has a peculiar "ring chromosome" that sometimes generates large-scale genetic rearrangements which produce offspring that cannot cross with the parent plant. This was "instant speciation," without any apparent action of natural selection or any slow accumulation of differences. It was an unfortunate case of erecting an entire theory out of a single bizarre case. Based on this one case, mutation pressure was claimed to be the critical evolutionary force, with new species arising when sufficiently large mutational jumps occurred. Ironically, this whole process amounts to a kind of materialistic Special Creation, lacking only a deity.

While the theories of the Mutationists were both specious in logic and spurious in experimental foundation, they did at least focus the attention of biologists on mutations which arose without any direct environmental determination. Moreover, De Vries supposed that mutations arose without any particular adaptive quality; they could be harmful or deleterious with respect to a given attribute. Thus, they invoked a "blind" evolutionary process, in which there was no necessary benefit. This was a high-water mark in the materialism of evolutionary biology. Unsurprisingly, the Mutationist geneticists and the neo-Lamarckians of the early twentieth century, with their almost spiritualist thinking, were fierce opponents. The victory of the former where heredity was concerned led to the abandonment of the inheritance of acquired characters. To some extent, this was perceived as a victory of Mutationism over Darwinism, because the latter was associated with such inheritance of acquired characters. This exacerbated the erosion of Darwinism engendered by the arrival of Mendelism itself, at least where scientific politics were concerned.

A major conceptual problem was also generated. For DeVries, who introduced the term into biology, a "mutation" was an abrupt genetic discontinuity, and by his definition, this necessarily generated a new species. Therefore, mutations were by definition the force giving rise to speciation. It took some time for this confusion to be purged from the literature, so that mutation and speciation were clearly separated.

One important development in the resolution of these issues was that the genetic system of *Oenothera* was being sorted out experimentally. It became apparent that it possessed a bizarre system of inheritance, essentially unknown in other species, which made it an inappropriate system for testing general evolutionary hypotheses.

Beginning in the 1910s, the future Nobel laureate H. J. Muller began to work on lethal mutations using the fruit fly *Drosophila*'s abundant resources of unusual genetic strains. In particular, Muller was able to treat these flies genetically so that they accumulated recessive lethal mutations. A particularly important feature of Muller's work was that he showed that such mutation rates increase when temperature is raised or when the fruit flies were subjected to radiation. This work convinced many that mutation was a conventional chemical process, depending on the provision of sufficient energy for activation of some chemical reaction(s). This energy could come from the kinetic energy supplied by elevated temperature, or it could come from the atomic particles arising from irradiation. In addition, the irradiation technique made it possible to increase mutation rates to levels much higher than had previously been obtained under normal conditions. In effect, if the experimenter wanted abundant mutations, all that was necessary was an X-ray machine. But those who watch horror movies late at night will not need to be convinced of this.

Later work by others in the 1940s also showed that such chemical compounds as mustard gas and formaldehyde could give rise to elevated mutation rates. By this point, there could no longer be any doubt of three basic points: (*i*) mutations are the result of conventional biochemical processes, rather than results of "will" or any other mechanism compatible with a neo-Lamarckian interpretation; (*ii*) mutations do not necessarily give rise to new species; and (*iii*) mutations do not have any tendency to be beneficial.

All of these points together undermined *both* neo-Lamarckian and Mutationist views concerning the sources of variation, and so those theories gradually crumbled. Their central flaw was that they

required mechanisms of heredity that were altogether too fortu-
itous, too hopeful. That either parental physiology or heavy radiation might produce novel adaptations is essentially a comic book
idea. Too many things have to go right at the same time. The end-
result of early twentieth-century genetic research was that no theory which explained evolution in terms of the *initial* supply of
genetic variants alone was left with any credibility. And the only
theory on offer which did not require directed mutation to make
evolution work was Darwinism.

POPULATION GENETICS: MENDELISM MEETS DARWINISM

In part, the fight between biometrics and genetics was
camouflage for the fight between Bateson and Pearson for power,
influence, and prestige within the British scientific establishment. It
suited both to agree that Mendelism and biometrics were incompatible, since they felt incompatible with each other. But there were
those who realized almost immediately that there was no necessary
opposition between Mendelism and biometrics. The first to publish
such a view was Udney Yule, a British mathematician. He developed an analysis of Mendelian loci with dominance which led to
correlation between relatives of the type that biometricians studied.
Problems with exactly fitting known biometrical correlations between the traits of parents and offspring arose, however. Yule suggested that allowing incomplete dominance and some degree of environmental variation could successfully account for these. In this
he was correct. In addition, Yule attacked the view that discontinuous patterns of inheritance implied discontinuous evolution. Instead, he suggested, continuous variation could be due to many
genes each of small effect, with selection acting to change gene frequencies at many genetic loci. Unfortunately, Yule did not make
enough of a nuisance of himself to fundamentally change the battle
lines between Mendelians and biometricians. A German physician
by the name of Wilhelm Weinberg performed similar calculations
in 1908, but he too was easy for both sides to ignore. A decade
elapsed before someone of sufficient force of character would publish on the problem.

Ronald Aylmer Fisher had the advantages of being young, determined, and brilliant. He also had all the benefits of a "public" school education at Harrow (the English school where Lindsay Anderson's damning *If . . .* was filmed) followed by a scholarship degree at Cambridge. While still an undergraduate, he began to work on problems in mathematical statistics. He published his first paper at the age of twenty-two in Pearson's journal *Biometrika*. At first he was under Pearson's influence, but later he adopted Mendelian beliefs. As Fisher's work began to indicate more and more support for a genetic analysis of quantitative characters, Pearson became progressively more hostile. Finally, Pearson rejected some of Fisher's work for *Biometrika*, and the hostility between them grew without limit.

In 1916, Fisher completed an analysis of inheritance which synthesized biometrics and Mendelism, essentially along the lines suggested by Udney Yule, but Fisher's analysis was far more extensive.[8] This synthesis provided the foundations of "population genetics," the hard core of contemporary Darwinism. Pearson did everything in his power to prevent this paper from being published, but it finally appeared in a relatively obscure Scottish journal. From that point on, Fisher could not be stopped. Here Pearson may have lost in part because he had met his match. Fisher was zealous and intolerant, while at the same time his mind was cold and mathematical. In other words, Fisher would do anything to win. And by and large, he did; where Yule and Weinberg had failed to be heard over the two feuding parties of Pearson and Bateson, Fisher managed to prevail. The fight between Mendelism and biometrics was over, even if each side still hated the other. It had been a fight over nothing, because Mendelism and Darwinism were fully compatible, leaving aside Darwin's insistence on continuous inheritance. It was perhaps fitting that the struggle had been played out against the backdrop of one of the most absurd catastrophes of history, the period leading up to WWI.

A TASTE OF POPULATION GENETICS

In 1908, the English Mendelian R. C. Punnett gave an address on "Mendelian Heredity in Man." Udney Yule was in the audience and made some comments about the effects of repeated crossing of two hybrids on the frequency of two alleles at a locus. This was a subject which Udney Yule and the American William Castle had published notes on, and it remained without a general solution. Punnett had, however, played cricket in school with G. H. Hardy, then one of the greatest living English mathematicians. He put the problem to Hardy and within a few weeks Hardy had worked out the "Hardy-Weinberg equilibrium," as it was later known, because Weinberg had independently and simultaneously obtained the same result. This mathematically trivial result was to be an embarrassment to the great Hardy, but despite that it is a cornerstone of population genetics theory.

This result is the simplest in population genetics, but it illustrates many of the most important features of the field. It assumes that there are a fixed number of alleles, or types of gene, at a locus, or genetic site, with random mating and no other evolutionary forces. The most important point to grasp about this theoretical situation is that genes are combined at random, and parents are combined at random to form mating couples. Everything that happens in mating and the production of offspring is analogous to shuffling a deck of cards for a card game. If there is no cheating, and no cards fall on the floor, then each hand will be drawn from the same number of cards. In bridge, for example, there will always be fifty-two cards dealt out in four hands of thirteen cards each. All of these cards are the same, every hand, it's just their arrangement that varies. Likewise, the Hardy-Weinberg Law shows that random mating preserves the allele frequencies, unaltered, from generation to generation. One of the important features of this result is that it also shows that Mendelian variation is *conserved* from generation to generation. There is no tendency for the variation to be lost, unlike the situation with blending inheritance, which we have already examined. An essential aspect of this conservation is that there is no directional trend imposed by Mendelian genetics. Neither allele is favored. Mendelian genetics, by itself, is not a force for evolutionary change. Thus, Mendelism is not a complete foundation for evolutionary theory. It is instead complementary to Darwin's scientific contribution. Mendel's work patched up the major holes left by Darwin's failure to solve the riddle of inheritance.

There is a summary statement that can be made. There is a lot of heritable variation affecting almost all characteristics of living things. This variation is almost always inherited according to the laws of genetics, with essentially no inheritance of acquired characters. The ultimate source of this variation is mutation, which is not directed according to the adaptive needs of the organism. The magnitude of mutational effects may be large or small, but whatever it is, it does not necessarily lead to the formation of new species. There is no directional trend arising from Mendelian genetics, as such. Accordingly, while there is genetic variation out there, its transmission from one generation to the next, by itself, does not explain evolution.

3 SELECTION

Nature Red in Tooth and Claw

AND SO, inevitably, we turn to the idea of natural selection, an idea that Darwin finally made intellectually unavoidable. If Darwin's role in the study of heredity was to set the stage for Mendel and the geneticists, with respect to the concept of selection he did far more. Indeed, much modern scientific reasoning about selection remains very close to the level that Darwin reached within his lifetime.

The greatness of this achievement can be understood best in terms of the historical leap that Darwin's work was. In all of human history before the publication of the *Origin of Species* in 1859, the concept of biological selection was only a flickering flame. It appeared intermittently at the high points of natural philosophy: classical Greece, perhaps some medieval Arab writings, and Europe after the Renaissance. But it played little role in the work of biologists or their colleagues.

In the hands of Darwin, the concept of selection became one of the most powerful in all of biology. Numerous problems and patterns of biology were suddenly seen as explicable in terms of natural selection. A considerable fraction of the issues that consume the attention of evolutionary biologists were brought forward by the work of just one man.

Ever since Darwin we have been gradually expanding on his basic theme of selection, integrating it with our knowledge of genetics, and forging experimental and other empirical approaches to testing theories of selection. To be an evolutionary biologist is to live in the shadow of Darwin's achievements, one of which was his theory of natural selection.

EARLY IDEAS OF SELECTION

The theory that a Platonic essence underlies each species, discussed in chapter 2, is compatible with some features of natural selection. After all, classical Greek biology had to account for the preservation of the particular essence, or *eidos*, of each particular group of living things. This problem may be solved in one of three ways.

First, it may be supposed that there is some divine force which continually restores or preserves an ordained pattern. Presumably this sort of intuition sustained the medieval European belief in the fixity of species. This type of explanation could not have been satisfactory to the classical Greeks, with their analytical bent, nor to the scholars of the Enlightenment in the eighteenth century, with their atheistic bent. But important Tory English biologists, like Richard Owen, still embraced this type of thinking well into the nineteenth century.

A second alternative is to deny the possibility of real variation. Apparent and sometimes monstrous variation is then dismissed as misleading. The real causes of an organism's "nature" are held to be below the surface, in a hidden *eidos* which does not vary. This is the Platonic solution. All the research of modern genetics has expunged this hidden essence, this "Ghost in the Machine." Naturally occurring genetic variation is abundant and close to omnidirectional. Mutant animals can vary in eye color and limb number. They may even lack eyes altogether. Thus, genetic variation is not constrained to preserve the fixity of species. This we have seen in chapter 2.

There is a third alternative which is relatively cunning. So cunning that people keep rediscovering it every thirty or forty years, and offering it to the world as a great revelation. One of its earliest known proponents was Aristotle, the man who really founded biology. Aristotle allowed the possibility that there could be inherited variation of real effect, but then suggested that whatever variant does not function efficiently will perish, and not be preserved. This notion of not being preserved must reflect an absence of descendants, and thus natural selection of a sort. In effect, this is purifying or elimination selection, with the variant forms being "eliminated" and species thereby "purified." This will result in the preservation of the favored type, which can be naturally associated with the Platonic ideal for the organism, at least for someone with an education steeped in Greek and Latin texts.

This elimination theory has been offered by a wide variety of writers other than Aristotle, from Empedocles through Lucretius, in classical times. There was also a flurry of Enlightenment writers who rediscovered the idea, among them Diderot, Rousseau, and Hume. Since the biological sophistication of these authors was typically quite low (they tended to regard themselves as being in the business of explaining everything), they don't constitute an improvement on Aristotle, whose knowledge of biology was immense, by comparison.[1]

WAITING FOR DARWIN

Evolution by natural selection was one of those ideas that was heralded by much foreshadowing. For example, in 1813 the Royal Society of England was treated to "An Account of a White Female, Part of Whose Skin Resembles that of a Negro," a paper delivered by William Wells, a member of the Society. Here, Wells points out firstly that there is abundant variation that can be inherited and that this variation has been used by animal breeders to create domesticated breeds. Then he goes on to suggest that a similar process could occur in man, whereby in Africa a dark-colored race must have had an advantage in bearing "the diseases of the country," so that "this race would consequently multiply, while the others would decrease. . . ." In this way, selection would create a new variety of man. This idea is clearly more of a *creative* natural selection. But it was largely ignored, and left undeveloped by Wells.

In an extreme case of self-defeating publication, Patrick Matthew added an appendix on natural selection to his 1831 book *Naval Timber and Arboriculture*. This appendix was ignored by virtually everyone until after Darwin had published his *Origin of Species*, so that its value as a basis for priority is purely nominal. Nonetheless, Matthew proposed in 1831 that selection can achieve active change in species by the utilization of variation. Furthermore, he suggested that species could evolve from each other as a result of this process of selection. So, in only a few pages, Matthew gives a somewhat vague anticipation of everything that Darwin was to work on from 1837 to 1859. Even the style of argument is similar to Darwin's.

It is an important question to ponder why these earlier writers did not give evolutionary biology its starting point, the way that Darwin did. One reason is simply that they did not *develop* the ideas enough. Natural selection has many intricacies, particularly if it is used to account for evolution. Once he thought of natural selection,

Darwin felt that at last he had "a theory by which to work." In other words, the essential idea was only the starting point. In science, it is not enough to have a bright idea, or even several bright ideas. It is necessary to develop a cogent, articulated body of theory in which both implications and weaknesses are addressed. We now turn to the development of just such a body of theory, Darwin's theory of evolution by natural selection.

DARWIN'S CREATIVE NATURAL SELECTION

The starting point for Darwinism is ecology, the study of the interactions between species, within species, and between species and environment. With Malthus, Darwin was a founder of ecology, as well as evolutionary biology. There are few better introductions to ecological principles than the third chapter of the *Origin*, "Struggle for Existence." In this short essay, Darwin covers all the basic elements that would later grow into ecology in the twentieth century. Of particular importance for Darwin, following the thoughts of Malthus, was the opposition of the abundant fertility of organisms to the many factors that can kill their progeny before they reproduce. "Hence, as many more individuals are produced than can possibly survive, there must in every case be a struggle for existence, either one individual with another of the same species, or with the individuals of distinct species, or with the physical conditions of life."

The next point in Darwin's ecological analysis is that mortality imposed on the reproductive surplus will not always be at random. Indeed, we have already seen that there is abundant genetic variation affecting all sorts of characters. Some of this variation will affect resistance to disease, survival under starvation, and success in combat for mates. One historical example of this is the strong selection for resistance to disease that occurred during the repeated plagues of medieval Europe. Mortality levels during these epidemics could rise to one-third or one-half of a local population, as discussed in chapter 6. If an organism has enhanced attributes, where survival and reproduction are concerned, then its contribution to the next generation is likely to be larger. Conversely, if an organism is deficient in any way which affects survival or reproduction, then it will contribute less. And if the basis of this difference can be inherited, then some effect on the next generation is inevitable. Darwin's conjecture was that the ability to survive and reproduce would thereby be enhanced, and the attributes associated with this would spread.

Natural selection is not, however, very rapid. Its basic component mechanisms are limited by unavoidable features of ecology and genetics. The total amount of mortality imposed by selection in any one generation cannot be too large, or the population will die out. The amount of available hereditary variation in any one generation may be limited, simply because of a limit to the number of different variants in a population of finite size. Therefore, it is unreasonable to expect natural selection to work dramatic changes in a single generation. Indeed, while it is important for anyone learning about evolution to perceive the potential power of natural selection, it is also very important to understand that natural selection is usually weak in its impact on any one generation. Indeed, it will often be difficult to detect its effects over a single generation. The fact that it is rarely seen to have much effect on a short time-frame suggests that its action cannot be instantaneous. And so Darwin always emphasized, in gradualist fashion, that the action of natural selection would be cumulative over many generations. Slight, almost imperceptible modifications would spread through species, these modifications becoming large in scale only over many thousands of generations.

But despite this caution, Darwin also argued that natural selection was the primary driving, adapting force in the history of life. For example, Darwin held that the differences which natural selection could accumulate over great periods of time would be sufficient to explain the differences between species. Thus, in Darwin's formulation, the seemingly *qualitative* differences between species are the result of merely *quantitative* changes amplified over many generations. Still greater differences between large taxonomic groups, like insects and mammals, Darwin held to be due to the still more prolonged action of natural selection producing divergence. No additional evolutionary forces are required; natural selection is sufficient. This is not to say that Darwin dogmatically excluded other potential evolutionary factors. He did in fact invoke a wide variety of effects of environment, use, disuse, and so on. But for him natural selection was a sufficient platform on which to erect a materialistic and mechanistic account for life. At no point in this scheme is there any need for leaps of biological innovation, nor is there any need for an intervening deity. It is entirely a process of stepwise accumulation of small relative improvements, with no ultimate end or vision, only the long-term effect of differential rates of net reproduction.

DARWIN'S MARSHALING OF EVIDENCE
AND ARGUMENT

At no point in his career did Darwin have any examples of natural selection demonstrably operating on a specific population in the wild. We have examples of this kind now, which will be discussed shortly. But Darwin had to support his theory using less direct evidence.

The best point in favor of Darwin's theory was the power of artificial selection to create new animal and plant breeds. Anyone who has seen a Saint Bernard beside a chihuahua, products of deliberate human breeding coupled with accidents of husbandry, should believe in the power of selection. The point of raising artificial selection in support of natural selection is that it shows the power of selection to create organic change. It does not show that selection occurs in nature, which is another problem. For that Darwin used the basically ecological arguments of Malthus.

A host of problems arise in thinking about natural selection as a mechanism for the generation of the diversity and adaptedness of organisms. Darwin himself anticipated most of them, and developed counterarguments.[2] One problem is that there are no apparent intermediates between many species, yet gradual accumulation of differences during evolution by natural selection must allow the possibility of many intermediate forms. Why don't we still see these intermediate forms? Darwin's rebuttal was that the later products of natural selection would have eliminated the less completely developed intermediate forms. An obvious example of this is that *Homo sapiens* is the only surviving hominid species, when once there were at least several. Directly or indirectly, the other hominid forms have been extinguished.

Another problem is the seeming improbability of selective transitions between radically different ways of life, such as that between terrestrial and aquatic mammals. Related to this is the existence of organs of extreme perfection, partial versions of which would seem useless. The type specimen here is the vertebrate eye, the components of which would seem to be of little value on their own. To counter these arguments, Darwin analyzed individual cases which seem difficult. For example, he pointed out that all nervous tissue is somewhat sensitive to light, and furthermore there are many types of primitive eye known among animals, which nonetheless provide photo-detection functions. It is far from the case that the binocular color vision of primates is the only useful form of vision. Even primitive forms of light detection can be useful to simple organisms that might then be able to detect a looming predator by the shadow that it casts. In addition, Darwin points out that transitional structures can serve functions different from their final functions. Thus, some fish use their swim bladder for some degree of air respiration, even though the swim bladder was not originally a primitive lung. In fact, Darwin's suggestion that the swim bladder is evolutionarily related to the terrestrial vertebrate lung is now supported by overwhelming evidence. Organs do not necessarily evolve in any simple fashion toward some ideally assembled, characteristic form.

As a Darwinian would expect, organs evolve by meandering evolutionary pathways in which particular functions may be acquired over some hundred million years and then be lost over the next hundred million years. Even if, or especially if, our limited human intuition finds it difficult to conceive what evolution is up to, there is little reason to suppose that the complexity or obscurity of the evolutionary story suggests that some entirely different theory, like creationism or continuing cosmic guidance, is more likely. Rather, Darwinism can readily explain the broad features of evolution, while expecting it to be perverse and indirect in its action, unlike theories of life that invoke supernatural teleology.

Early Selection Experiments

It was in the early part of the twentieth century, with the controversies engendered by the rediscovery of Mendelism, that experimental tests of Darwinian theories of selection were first performed. Initially, the experimental results were grim for Darwin. Several high-profile experiments gave results that were difficult to interpret in terms of selection. We now know that these initial problems were caused by bad experimental designs. But it was only a matter of time before a laboratory would perform experimental work of sufficient quality to subject Darwinism to an appropriate test.

That laboratory belonged to William Ernest Castle, chair of genetics at Harvard University in the first part of the twentieth century. Castle was a highly influential mammalian geneticist, who produced many of the most important American doctoral students of the next generation. (Among his students was Sewall Wright, the greatest American evolutionary biologist.) At first, Castle was very much a disciple of William Bateson, holding a Mutationist interpretation of the mechanism of evolutionary change. It certainly was not the case that Castle experimentally tested Darwinism in order to support it. If anything, he wished to destroy it.

The key experiments involved selection on coat pattern in hooded rats. These rats have a dark stripe on their backs, starting from the head. This "hood" varies in size, from rat to rat. Castle and his assistants selected for rats with large stripes in one line and selected for rats with smaller stripes in another line. Both lines responded to selection, and the response to selection was stable. That is, they produced some rat lines with thicker hoods and some with narrower hoods. But in an unfortunate blunder, Castle went all the way back to the original Darwinian point of view and suggested that the selective response was due in part to blending inheritance. There was no substantial evidence for this; it was merely a case of scientists sticking with conventional ways of putting ideas together. The idea of Darwinism together with Mendelism was still difficult for biologists to accept. Fortunately, the work of Sewall Wright and others on the genetics of coat color in these rats and in other mammals showed that the coat patterns were not inherited by blending inheritance. Instead, Mendelian inheritance was clearly indicated. Experimentally, Mendelism and Darwinism could work together.

In any case, Castle had shown clearly that selection could progressively accumulate differences between lines, the essential point for Darwinism. Other selection experiments, such as those of Edward Murray East with corn oil content from 1910 to 1918, also showed that selection could be a powerful force for cumulative change. The resultant situation was one in which both Mendelian inheritance and Darwinian selection were clearly established experimentally. The outstanding problem was how to explain their compatibility.

Resolution of Mendelism and Darwinism

From 1910 to 1920, many factors pulled together to win the day for an evolutionary biology which combined Mendelism and Darwinism, but lacked neo-Lamarckism, Mutationism, or blending inheritance.[3] The issue which was crucial was the nature of "continuous variation," such as that of human height, particularly its genetic basis, if any. We have already mentioned R. A. Fisher's theoretical analysis. As important was experimental work with multiple genes. In cereals, it was shown that some patterns of inheritance required the supposition of multiple genes affecting the same character, not just a single gene for each character. This was very close to the model that Fisher had developed mathematically.

In addition, by about 1912 the *Drosophila* laboratory of T. H. Morgan had begun to isolate Mendelian alleles of small effect. Further work with *Drosophila* was turning up abundant evidence that eye color was subject to variation involving at least eight loci, supporting the notion of multiple factors. This established the plausibility of explaining all "continuous variation" in terms of Mendelian factors.

Perhaps the following statement from H. S. Jennings, once a disbeliever in Darwinism, best reveals the change that took place:[4]

> It appears to me that the work in Mendelism, and particularly the work on *Drosophila*, is supplying a complete foundation for evolution through the accumulation by selection of minute gradations. . . . The objections raised by the mutationists to gradual change through selection are breaking down as a result of the thoroughness of the mutationists' own studies. The positive contribution of these matters to the selection problem is to enable us to see the important role played by Mendelism in the effectiveness of selection.

Natural Selection in the Wild

But many critics of Darwinism have not been satisfied by research that shows the power of selection to produce genetic change in laboratory populations. Their concern is that this does not necessarily show that these same processes operate in nature. And there are good reasons for taking this criticism seriously.

A perennial topic for debate among population biologists is the relationship between information obtained in laboratory settings as opposed to that collected in the field. Most ecologists adopt the standpoint that processes which are only known in the laboratory are not therefore known to occur "in nature." Others, particularly those with a genetic bent, tend to reply that the laboratory is part of the natural world too, not a bizarre other-worldly realm.

In most of physics and chemistry, a laboratory demonstration is a perfectly valid one. Chemists feel no compulsion to repeat their reactions in forests during thunderstorms. Similarly, geneticists will show little interest in repeating their crosses out-of-doors. For all of these scientists, if it happens in the laboratory, and always happens the same way there, then there is little further information to be gained outside.

The legitimate problem that the ecologists raise is that many of the circumstances which exist "in nature" are destroyed by laboratory culture. There is no weather, no predators, and no competitors. In addition, some population geneticists point out that patterns of gene expression can change when organisms are brought from the field into the laboratory, rendering laboratory inferences, on their own, dubious.

Perhaps two basic points of view can be extracted from this welter of disparate opinions. The first is that laboratory experiments have the definite value of showing what can occur, not what does occur and certainly not what must occur. The experiments of Castle on selection in hooded rats showed that selection could accumulate Mendelian allele differences, so that Darwin's natural selection could be the mechanism of evolutionary change, rather than Mutationist saltations. Before that experiment, many doubted that the mechanism of selection could be effective in producing significant change in populations.

The second point, though, is that it remains to be shown that mechanisms which can operate in the laboratory in fact do operate in nature. Selection may be effective in the laboratory or on the farm. That does not prove that it is effective in nature. Such demonstrations are required, over and above laboratory research. These demonstrations have been supplied. A few examples will now be given.

One of the earliest demonstrations of natural selection in the wild was that of W.R.F. Weldon, using the crab *Carcinus moenas*.[5] What Weldon observed at first was a historical record of decreasing "frontal breadth" in crabs found near an estuary that was becoming more and more clogged with silt and sewage. The obvious inference is that smaller crabs can somehow survive this polluted environment better than larger ones. To test this, Weldon compared crabs subjected to salt water environments with and without silt and sewage. Those with smaller frontal breadths survived better in silted water. This fit well with the original inference of natural selection, indicating its action in the estuarine environment.

Perhaps the best documented example of natural selection in the wild is that of industrial melanism. Throughout the nineteenth century, extensive use of coal-fired furnaces in industrial plants led to the production of large quantities of soot in England. This soot was so extensive that many square miles of English countryside were substantially darkened, particularly the trees. Beginning in the middle of the nineteenth century, lepidopterists began to find more and more dark forms of butterflies and moths. More than a hundred Lepidopteran species have undergone this change in melanism. (Melanin is the term for darkening pigment.) This phenomenon, as such, is called "industrial melanism."[6]

The question is how this evolutionary change occurred. E. B. Ford has analyzed this situation as follows. These species have an important behavioral feature in common: they all rest upon tree trunks or rocks, deriving their protection from predation from resemblance to their background, be it trunk or rock. These forms lack noxious chemicals for dissuading predators. Nor do they exhibit cryptic behavior patterns, such as retreating into crevices. A particularly important aspect of the camouflage used by these species is the lichen which normally grows on tree trunks and rocks in the English countryside. This lichen is killed by pollution; soot then takes its place as the chief feature characterizing the resting surfaces.

One obvious hypothesis is that natural selection has favored melanic forms because they match the sooty tree trunks and rocks better. That in turn raises the question of how natural selection could favor this matching; what is the camouflage for? For a time, in the 1920s, it was proposed that pollution directly altered the moth physiology so as to lead to the production of more dark pigment. Lamarckian mechanisms could thus be invoked to explain the retention of dark pigment in laboratory-reared moths. In fact, careful experiments revealed no such Lamarckian effect. Moth coloration followed relatively simple patterns of inheritance, these patterns often indicating just one main Mendelian gene for dark coloration.

What now appears to be the predominant evolutionary mechanism for industrial melanism is natural selection arising from predation by birds. The demonstration of this was primarily the work of H.B.D. Kettlewell. The species that he particularly concentrated on was *Biston betularia*. In this case, the melanic form arises from either one of two dominant alleles, *carbonaria* and *insularia*, a classically Mendelian system. Some of Kettlewell's work involved the release of large numbers of moths of different types, and then observing differences in their predation rates. In nonindustrial areas, which still had light-colored lichen on tree trunks, of 190 moths captured by birds, 164 were *carbonaria* and 26 were light-colored, when equal numbers were released. In industrial areas, with their darkened vegetation, of 58 moths taken by birds, 43 were light-colored and 15 were *carbonaria*, again with equal numbers released. Other evidence points in the same direction, particularly recapture levels upon release in large numbers, *carbonaria* being retrieved in much higher levels in industrial areas. The industrial melanism story is one of the classics of natural selection. Particularly valuable is the fact that a number of moth species evolved industrial melanism independently, so that idiosyncratic features of a given species are not plausible explanations of the phenomenon.

Perhaps the best understood human case of natural selection is sickle-cell anemia. This is apparently one of several mechanisms by which human populations have adapted to the spread of malaria, a disease caused by a blood parasite that is potentially fatal. Sickle-cell anemia causes red blood cells to take on a sickled shape. Such deformed cells cause circulatory problems that are often fatal, but the genetic condition is common in sub-Saharan Africa, where malaria is common. The key to understanding the way selection works in this case is the fact that individuals with sickling are also more resistant to malaria. There are three genotypes. The normal one has no sickling, but is highly susceptible to malaria. With one gene for sickling, of the two at the genetic site, there is some minor circulatory impairment as well as resistance to malaria. With two copies of an allele for sickling, survival away from modern medical care is greatly reduced, regardless of malaria. In effect, sickling protects against malaria, but severe sickling kills. This is a situation where the combination of normal and sickling genes is the best. That is the favored type, but it can't be fixed by natural selection, because it isn't a pure type. So in this case selection acts to maintain genetic diversity, or "polymorphism." This type of selection is often called "balancing selection," to distinguish it from the type of selection that Darwin first envisioned, which is usually called "directional selection."

KIN SELECTION

One of the basic types of behavior is interactions between individuals, particularly those interactions that affect fitness. A characteristic observation of scientists studying mammalian behavior, such as that of wolf packs, is of one individual aiding another. Pairs of male chimpanzees will distract copulating pairs of males and females, apparently so that one of the two intervening chimps can sneak in and copulate with the female. Scrub jay fledglings will feed their younger siblings that have newly emerged from eggs. Honey bee workers toil ceaselessly rearing bee larvae, defending the hive, and finding food to bring back to the hive, yet they are sterile; none of the larvae being reared in the hive are their own progeny. This type of behavior is called "altruism." The term altruism in biology does not refer to the sentiments of animals. Rather, the term refers to actions which reduce the fitness of the organism performing them while enhancing the fitness of some recipient organism(s). Such behavior is an anomaly for evolutionary biology, since our nominal expectation is that natural selection should oppose any systematic tendency to behave in ways which reduce fitness. This puzzle has been solved, in part, by the development of *kin selection* theory.

All the basic ideas of kin selection can be understood by considering an unusual context, that of the behavior of identical twins. Suppose that these are chimpanzee twins, being observed in a large enclosure. Call them Abercrombie and Beauregard, A and B for short. Since Abercrombie and Beauregard are genetically identical, from the standpoint of natural selection they are equivalent too. If Abercrombie can give Beauregard ten bananas, which Beauregard can use to make three amorous seductions, and these bananas cost Abercrombie only one mating, then we might say that A gave B three matings, at a fitness cost of one mating. (For baboons, birds, and the like, you can substitute other behaviors by Abercrombie which will result in Beauregard having more success in mating.) The key to understanding this type of interaction is to bear in mind that A and B have the same genotype. When Abercrombie gave Beauregard the bananas, his act was altruistic in terms of its impact on his personal fitness, but it was beneficial, on average, for the joint genotype. The number of matings obtained from the ten bananas went from a prospective one to an actual three, a net gain of two total, or an average of one for each of the two animals. Under these conditions, selection can favor the evolution of the altruistic banana behavior.

But what we would like to know is the evolutionary criterion for the evolution of altruistic behavior when individuals are not genetically identical to each other. After all, genetic identity is relatively rare compared to lesser degrees of genetic relationship. Such other relationships include parent and offspring, siblings, and so on. In these cases, the genetic similarity is only partial. In 1964, W. D. Hamilton, a British evolutionary theorist and entomologist, proposed that the general criterion is that the product of the benefit of altruism and the degree of genetic relatedness be greater than the cost of altruism.[7] When this criterion is met, then it is possible for selection to favor altruism. It won't always favor altruism, but it may. This is a necessary condition then, not a sufficient one.

The tricky part in this criterion is the degree of genetic relatedness. For identical twins, this is one. They have full genetic identity. For parents and offspring, the degree of relatedness is usually one-half for each individual parent. This value arises because each parent contributes half of the genes to the offspring. For full siblings in mammalian, bird, and other species, the genetic relatedness is also one-half. For half-siblings, with only one parent in common, the genetic relatedness is one-quarter. For first cousins, it is one-eighth, and so on. These values probably account for J.B.S. Haldane's reply to the question of whether or not he would save a drowning man, if it meant losing his own life. Haldane replied that he would do so if he could save two brothers, four half-brothers, or eight first cousins.

These genetic relatedness questions also show that the reason why this type of selective context is referred to as kin selection is that it requires some degree of relatedness. In somewhat archaic

English, genetic relatedness means that there is kinship, and that the individuals involved are themselves kin.

INSECT SOCIETIES AND KIN SELECTION

A fundamental corollary of Hamilton's theory is that species with different patterns of genetic relatedness should have different patterns of behavioral organization. In particular, species in which individuals have higher levels of genetic relatedness should exhibit higher levels of altruism. That this is so is dramatically illustrated by the "social insects." The classic examples of social insects are to be found among the Order Hymenoptera, which includes honey bees, wasps, and ants. Members of this insect order have an unusual system of sex determination. Fertilized eggs become females, and have two copies of every chromosome, a condition called "diploid." Unfertilized eggs become males, which have only a single copy of each chromosome, which is called "haploid." This overall condition is called "haplo-diploidy."

The real importance of haplo-diploidy for the evolution of altruism is that it changes the structure of genetic relationships. Haplo-diploid full sisters, who have both a mother and a father in common, will normally have a coefficient of genetic relationship equal to three-fourths, as opposed to the normal one-half for siblings in a conventional species. The reason for this is that their father has only one set of chromosomes. Therefore all his sperm have this same set of chromosomes exactly, and all his daughters receive the same set. Immediately, this makes the genetic relatedness between sisters at least one-half. If the mother is in common too, then that increases genetic relatedness still further, up to a minimum of three-fourths. This is greater than the genetic relatedness of parent to offspring, which remains one-half. This means that a daughter gets a greater benefit from facilitating the production of sisters rather than daughters, all other things being equal. This can be done by becoming "mother's little helper" about the nest, rather than reproducing oneself. Therefore a basic prediction that we can make about haplo-diploid societies, when the females are diploid, is that they should frequently be organized into groups in which daughters help their mothers rear more daughters.

What about the males in this system? Interestingly, the brother-sister coefficient of genetic relationship is only one-fourth. This happens because the brothers never have paternal genes in common with their sisters, because they (the brothers) are produced from unfertilized eggs. In fact, this genetic relationship is the same as that for half-siblings, which is really what the brother-sister relationship is in haplo-diploid species. Things are even more dramatically different if we look at the genetic relationship between fathers and sons. Since males are produced from unfertilized eggs, there is no genetic relatedness between fathers and sons. Thus the males are *less* related to other members of the family group in haplo-diploid systems compared to conventional diploid systems. The corollary of this is that males should exhibit little tendency to altruistic behavior, striving mostly to mate, rarely assisting their sisters, and never helping new male progeny.

The overall picture then is one of the mother who produces numerous offspring, assisted by nonreproducing daughters, with males of little use, except when it comes to mating. And this is exactly what occurs in some of the Hymenoptera. Their "societies" are centered around a queen, who produces vast quantities of eggs, is enlarged, and does little in the way of active food gathering. The gathering of food, construction of hives, nests, and colonies, defense of the nest, and so on are all taken care of by sterile female workers. Sometimes, especially in the case of ants, these females are structurally modified to specific tasks such as fighting, having greatly enlarged mandibles and head capsules. The males, meanwhile, are not workers. They exist, in low numbers, solely in order to mate with the queen. For this reason, they are normally referred to as drones. All told, there is a striking correspondence between the most organized Hymenopteran societies and the predictions of Hamilton's theory.

An important test of this theory arises in social insects which are *not* from haplo-diploid groups. In conventional genetic systems, there is symmetry between mothers and fathers as well as brothers and sisters. In such societies, where Hamilton's theory is concerned, we expect that both males and females will evolve altruism to an equal extent. Thus, if there is a worker caste in such societies, it should have both males and females. In addition, rather than a solitary queen, there should also be a reigning king. This pattern is found among termites, particularly the "higher termites" of the family Termitidae. These insects build large mounds, where each mound is normally built by the progeny of just one mated pair of termites. The mound is started by the queen, sometimes assisted by the king. Then there is a first generation of workers, with subsequent generations having a mixture of workers and soldiers. Except when reproductively competent termites are let out, the mound normally remains closed to the surface. All feeding takes place underground. Both sons and daughters are workers, the male workers of some species being larger than the female workers, other species exhibiting the reverse pattern. Nonetheless, there is no fundamental division of role between males and females in these societies; both sexes are "monarchs" and both sexes are workers. This beautifully fits the predictions of kin selection theory.

It may be of interest to note in passing that there is a subterranean rodent, the naked mole rat of Africa, which exhibits an almost complete mammalian parallel with the termite. It too has adapted to feeding on underground woody structures, like roots, and exhibits high degrees of altruism. There is a dominant male and female couple which have numerous nonreproductive progeny. The progeny perform worker roles, gathering food and defending the labyrinthine burrow. This is the only known case of such caste-like social organization among mammals. Since we mammals are genetically conventional, the mole rat follows the termite pattern in its lack of sexual differentiation of role, again fitting the expectations of kin selection theory.

Strategy Selection

Sometimes the value of an adaptation depends on the adaptations of the other individuals within the population. For example, if you are the only individual with large fangs in a population, then they can be used to demolish your rivals in combat. But if everybody has the same fangs, their value is considerably reduced. You might yourself be killed in the course of unrestricted combat. Problems like these are called *evolutionary games*, where the idea is that there is a contest between members of the same species for some item or outcome which will increase their fitness. An example of an evolutionary game is birds contesting feeding territories, where such territories may be patches of field, individual trees, or lengths of riverbank. Another example is when stags compete with each other for female deer in the mating season, bellowing and butting heads. In the first example, the stake was food, in the second it was mating. Either is obviously of benefit to fitness.

Game theory has its roots primarily within mathematics and economics. It is natural for an economist to wonder about complex strategies in a competitive situation, where the outcome of a behavior depends on what others do, because the economic behavior involved is that of a pretty smart organism, a human. A distinctive body of theory has been developed for such human strategizing. With animals, the application of game theory wasn't as clear at first. The key breakthrough came when W. D. Hamilton, John Maynard Smith, and George Price had the idea of unbeatable strategies, strategies that could not be improved on if all members of a population adopt them. The unbeatable strategy of a game is expected to be an end-point for evolution, because once we get there we can't go anywhere else. In other words, unbeatable strategies are our initial predictions for the strategies that we expect evolution to produce in the organism under study.[8]

Perhaps an example will make the concept clearer. One of the puzzles of animal behavior is that it is so often lacking in aggression when we would expect violence to be common. For example, contests between males for females would seem to be cases where extreme aggression should prevail. The reality is different. Deer are equipped with antlers which could do a lot of damage if stags attacked each other from the side or rear. Instead of doing so, they circle each other so that they face each other directly, and then engage in antler-to-antler attacks. These spirited confrontations are rarely more than shoving matches. With these rules of combat, the chances of severe injury to the stags is slight. Evolutionary game theory turns out to provide natural ways of explaining such limited aggression. Let us suppose that there are three possible moves in a stag contest: Display (D), Escalate or attack (E), and Retreat (R). Injuries may result when the opponent attacks (E), resulting in forced retreat (R) and an injury cost. Let's assume that the benefit of mating is less than the cost of injury. If one stag retreats (R), with or without injury, the other stag mates. Displaying interminably will have some time-cost, but this is probably much less in magnitude than the cost of injury or the benefits of the prize. A single contest sequence might look like this, time advancing from left to right:

Stag 1: D D D D D R (injured)
Stag 2: D D D E E

In the end, Stag 1 is injured, while Stag 2 gets the mate. Stag 1 was the loser, and Stag 2 was the winner, from a Darwinian standpoint. Consider two basic, alternative strategies for playing this game.

HAWK: Always plays E, until victor or vanquished.
DOVE: Always plays D, until opponent plays E, in which case it plays R immediately, escaping injury. We assume that two Doves eventually cease Displaying and one cedes the resource to the other, with each having an even chance of winning.

To analyze the evolutionary dynamics of this situation, we have to consider the four different contests that are possible with these alternative strategies. When Hawk plays Dove, Hawk will always win. Dove won't be injured, as it will retreat first. So Hawk gets the evolutionary benefit of the mate, while Dove will receive nothing. When Dove plays Dove, after some jumping around, one of the Doves will win, and one will lose. Both will pay the price of their protracted displays, and one will receive the benefits that go with the prize. When Hawk plays Hawk, things become more extreme. One of them will be injured in the Escalated contest, at great cost. The other will luckily escape injury, and so mate.

Pay careful attention to the *average* benefits received. A Hawk playing against a Dove receives, on average, the fitness enhancement of the mating, which is positive. In this same contest, against a Hawk, a Dove receives nothing. Dove against Dove receives, on average, half the value of the mating, because only one of the Doves will win, minus the cost of protracted display. The net winning will probably be positive, but nothing as great as the benefit that a Hawk receives against a Dove. Things become most interesting in the Hawk versus Hawk case. Half the time a particular Hawk will win, and half the time that Hawk will be injured. Because the cost of injury is greater than the benefit of mating, the average payoff in a Hawk-Hawk contest is negative. Hawks are actually behaving self-destructively, on average, by being so aggressive toward each other.

We can conclude from this that neither Hawk nor Dove is an unbeatable strategy. The reason for this is that each can invade populations composed only of the other type of strategist. Consider a population composed entirely of Doves. They are settling their contests amicably with an average winning that is positive. A sole Hawk invading an all-Dove population has contests only with Doves, always winning more than what the Doves get. So it has greater fitness and multiplies. That's fairly obvious. What is more subtle is the fact that a Dove can invade an all-Hawk population. The Hawks are beating up on each other, with an average net loss of fitness. A Dove comes wandering into this population, sees the mayhem, and retreats from each and every contest, before injury. It then receives nothing. But note that this is more than the Hawks are receiving, on average.

The bottom line on this is that neither simple pacifism nor outright militancy seems to win the day in animal contests. What does? Consider the following possibility.

RETALIATOR: Play D, unless opponent plays E, in which case play E.

Consider the prospects for Retaliator against Hawk. Retaliator is just a Hawk in all-Hawk populations, so it can always invade them. But Hawk does not do as well as Retaliator in mixed populations of Hawks and Retaliators, because the Retaliators are Dove-like in their interactions with each other. The more Retaliators there are, the more Retaliators act like Doves, and the better Retaliator does relative to Hawk. This will tend to force Hawk out of the population by straightforward individual selection. Once Hawk is gone, Retaliators act just like Doves, so that seemingly they are pacifist. But they are always ready to fight, upon provocation.

The scientific significance of an analysis like this is that we expect armed truce to prevail in the animal kingdom. There are a number of cases in which this has been shown to be true. For example, rhesus monkeys have ritualized combat, in which the loser accepts harmless incisor bites to conclude the combat. If, at the point when incisor bites are appropriate, the opponent instead bites with its canines, which are much more dangerous, then the bitten monkey will fight back viciously. The canine bites are perceived as escalation, which in turn elicits escalation in the responding monkey. They appear to be using the Retaliator strategy. Indeed, this type of behavior appears to be ubiquitous in the animal kingdom.

TERRITORIALITY

We have already seen how strategy selection can be used to explain the evolution of limited aggression. Another type of behavior that can be explained in terms of strategy selection is territoriality, in which one animal is apparently willing to cede possession of a particular area, an item of food, or a mate just because some other animal got there first. This type of behavior is widespread, not just among intelligent animals like birds and mammals, but even among butterflies and other insects. Again, it seems difficult to account for this type of behavior on the basis of individual selection. Why should the second animal accept this concept of ownership? Evidently animals in nature suffer from the depredations of bourgeois individualism.

Consider the following strategy:

BOURGEOIS: If there first, play Hawk; if second, Retreat.

Notice that the payoff to Bourgeois against Bourgeois is half the value of the contested item. There is never any fighting, or risk of injury. There isn't even any time wasted on display. Whichever one gets there first wins, the other retreating without injury. And since, on average, any animal has an equal likelihood of getting there first, the benefits are distributed widely through the population. This seems like a kind of utopian arrangement.

But all utopias have defectors. What about the local anarchist, who pays no heed to property rights or other such oppressive notions? Let us reintroduce the Hawk, and suppose that the Hawk strategist will always attack in order to get whatever type of property or territory it wants, ignoring who arrived first. If it happens that Hawk arrives first, then the Bourgeois will cede the prize immediately. Under these conditions, Bourgeois is just like Dove. If Hawk arrives second, Bourgeois is just like Retaliator, and thus another Hawk. An escalated contest will occur, won by Hawk half the time, Bourgeois the other half.

Bourgeois is an unbeatable strategy because its payoff against itself is greater than the payoff received by invading Hawks. If Hawks are rare invaders, they will play against Bourgeois only, and Bourgeois likewise will be playing against Bourgeois almost exclusively. The payoff of Bourgeois against its own ilk is half the value of the prize. The payoff to Hawk against Bourgeois is this same value plus the net returns from its escalated contests with Bourgeois. But this latter quantity is *negative*, because the net returns from escalated contests are determined by the difference between the costs of injury and the benefits from winning, where the costs of injury are assumed to be greater. In effect, Hawks lose out because they pay the price of an aggressive lifestyle, compared to the peaceable Bourgeois strategist.

Bourgeois also does better than Dove, because it expends no time on display. All disputes about access are settled immediately on the basis of ownership. Thus, we see that Bourgeois is an unbeatable strategy relative to both Hawk and Dove. As a result, we expect territoriality with respect to food items, terrain, and females to be commonplace. Animals should frequently exhibit a sense of "property," such that the one that gets there first is ceded the property from then on. Cases where this behavioral pattern breaks down should lead to Hawk-Hawk types of encounters, with escalated violence.

It is not a good test of this model that territoriality is indeed common. It was the phenomenon to be explained in the first place. The key prediction of this model is that it suggests that if you can fool two animals into believing that they are the owner, then you should observe a particularly nasty fight. There are two quite different cases where this has been done.

The first was with the Hamadryas baboon, *Papio hamadryas*. This baboon is usually found in troops with only one mature male. This male has all the mating prerogatives. In a primate rearing facility, males of this baboon species were given alternate access to a particular female. Thus male A was given first access to this female, then male B was introduced into the pen. Male B accepted male A's "ownership" of the female, without challenging male A. Then male B was given first access to another female, and male A was introduced to the pen. Now male A ceded ownership to male B. This showed that some form of dominance relationship between the males, separate from the females, was *not* the determining factor in the interaction. But if a male who has had access to a female, or set of females, is removed from them, while another male is given access, without the first male seeing, upon reintroduction of the initial "owner" there is a major escalated fight. All of this behavior is in keeping with the Bourgeois strategy.

Our second example comes from a very different organism, one so simple that you would think that ownership would be beyond its neural processing. This organism is the swallowtail butterfly, *Papilio zelicaon*. This species is relatively rare, and therefore it too is faced with the perennial problem of finding a mate. This problem has been largely solved by "hilltopping": males go to the tops of hills and females fly uphill when they are ready to mate. (That's any adult virgin, in this species.) Thus, hills become limiting commodities for males; they all want to be at the tops of hills. Normal male behavior in fact corresponds to the overt Bourgeois strategy; first one there gets to keep the hilltop. An experiment was performed in which two male butterflies were allowed to take up residence on a hilltop on alternate days, being kept covered and away from the hill otherwise. This way, on the Bourgeois model, both should infer that they are owners of that hilltop. When both were finally placed on the hilltop at the same time, they engaged in a protracted fight. Again this meets the predictions from the evolutionary game theory analysis; these are evidently Bourgeois butterflies.

INFERRING ADAPTATION

One of the troubling features of evolutionary biology is that selection is hard to observe. The cases of natural selection in the wild that have been presented here were unusual in being readily observable. The more general rule is that selection is so difficult to detect that nothing short of enormous efforts and considerable resources would suffice to detect it in any particular case. And this is more or less what Darwin expected.

But there are some ways around this problem. Evolutionary theory tells us that the average "fitness" of a population, also known as its average "rate of reproduction," tends to increase when natural selection is occurring. (In terms of individual organisms, fitness is usually given by the net reproductive output, once you allow for mortality before reproduction.) This isn't an absolute mathematical truth, but it is often true, in theory. If we find the fitness of a population increasing from generation to generation, and do not know anything else about the organism, the natural conclusion would be that selection is occurring. If at the same time, we notice the frequency of a particular attribute, such as darker coloration, increasing at the same time, it would also be reasonable to assume that dark coloration, or some attribute related to it, was being selected for. We have thereby indirectly inferred the action of selection.

Inferences like this can be made even less directly. Suppose we find that two populations of a species, inhabiting different regions, tend to have different body sizes. In particular, suppose that those inhabiting the colder regions are larger. It is common for biologists to infer that larger body size is an adaptation to a cold climate. (We could then preclude differential physiological responses to temperature by comparing body sizes of individuals from the two different populations after the same amount of exposure to low temperatures.) Thus, the appearance of adaptation is often used to infer the existence of a corresponding pattern of selection.

An entire style of research has grown up around this theme. Look for signs of morphology, function, or behavior which appear to be adaptations, and then look for mechanisms of selection which might have generated these adaptations. Examples of these selection mechanisms are: selection for resistance to some physical threat to survival, such as extreme heat or dehydration; selection for assistance to siblings; and selection for weaponry and tactics for combat with other males during mating season. Practically any physical attribute or function of an organism can be viewed as an adaptation to something, which is a problem.

If an evolutionary biologist comes upon a particular attribute of an organism for the first time, the natural temptation is to regard it as an adaptation and set about explaining it in terms of selection. Say that we are dealing with a spider weaving a web. A natural hypothesis is that it is there to catch prey. But it turns out, upon careful observation, that the spider does not catch prey this way. So that adaptive explanation has failed. Some evolutionary biologists, noticing dew on the web and the xeric conditions the spider lives in, might suggest that the web is instead an adaptation which reflects selection for water acquisition.

The question is, to what extent can we go on looking for selectionist ways to explain all the attributes of an organism? When we go on like this, our efforts illustrate the "adaptationist program." It is bitterly opposed by some evolutionary biologists as nonscientific, opposition that creationists delight in. If we suppose that biologists cannot seek endlessly for selectionist explanations, then what is the alternative? Do we fall back on God whenever the first naive guess does not work? The reality is that selection is not all-powerful. Fundamental constraints condition the evolutionary process, preventing selection from achieving perfection. These constraints are genetic and historical. We will discuss each in turn.

GENETIC LIMITS TO THE ACTION OF SELECTION

"Pleiotropy" is the glue and rubber bands of genetics, and one of the main causes of limits to the action of selection. But pleiotropy is a genetic phenomenon which shouldn't have a special name, because it is universal. Pleiotropy refers to multiple effects arising from a single genetic difference, as opposed to a single effect. From what we know of gene expression, from molecular to organismal levels, virtually all genes have multiple effects. The problem that pleiotropy poses for selection is that an allele (a particular model of a gene) may improve one attribute at the same time as it impairs another, relative to other alleles. The first effect could be selected for, while the second is selected against. If the second effect is large enough, the beneficial effects on the first character may not be realized, because the net effect on fitness is deleterious. If this genetic change is the only one which can enhance the first attribute, then evolution will be blocked from further improvements in that direction.

Consider as an example "alloparental" behavior.[9] This type of behavior occurs when adult animals "adopt" unrelated juveniles, feeding them, defending them from predators, and so on. In mammals, this occurs most often when adults are without progeny of their own. Such "adoption" is generally not successful, the adopted young often dying. Typically the adopting "mothers" are recently mature females. Among birds, there is a high frequency of "brood parasitism," in which adults lay eggs in the nests of others. Some species, such as cowbirds and cuckoos, exclusively reproduce by parasitizing the broods of other species. Since cuckoo nestlings, in particular, will go so far as to jettison the nestlings of the parents from the nest, the deleterious nature of such alloparental behavior is obvious. Why does it exist?

Female birds and mammals almost invariably make great investments of parental effort in the rearing of their young. Even some male birds and mammals do so. Much quantitative data shows how such parents lose weight and incur a greater risk of death by predation during rearing. There is intense selection for parental care behavior. Given this selection, there will be mechanisms which incline individuals of such species to receive positive (or "reward") stimuli from parent-offspring interactions. Such cues for behavior cannot be exact. While the parents of such species generally have well-developed offspring recognition abilities, they are not necessarily perfect. Furthermore, even if the adult does not perceive the juvenile as its own offspring, some of these "reinforcers" may still activate the parental-care pathways in the brain. (And that's why fluffy little puppies and kittens are "so cute!") The point is that there is selection for a general behavioral pattern, parental care of young, which elicits alloparental behavior which decreases fitness. The alloparental behavior is arguably a pleiotropic side effect of selection for the main adaptation. In this manner, selection can never produce perfection. But this is not a problem; selection is only a tinkering process of small relative improvements.

HISTORICAL LIMITS TO THE ACTION OF NATURAL SELECTION

Sometimes an analysis of an organism's adaptive needs suggests that it should have evolved an adaptation that it does not possess. An obvious example of this is respiration in aquatic mammals, like whales and dolphins. They should have evolved gills, yet they have retained air-breathing.

A general kind of adaptation which many organisms would benefit by is X-ray vision. From insects which parasitize hosts that live inside fruit to woodpeckers searching for grubs in trees, an adaptation that would enable an animal to see inside an opaque structure would be of obvious benefit. Yet no known organism has ever evolved any type of X-ray vision.

The reason for such failures of adaptation is that there is no genetic variability in the population which would allow the development of the structures required for such functions. There is no guarantee in evolution that such genetic variation must be present. When it isn't, selection cannot act.

It may be the case that an organism has some degree of genetic variability for an adaptation apparently required by the organism's ecological situation, yet there has been no response to selection. This is not necessarily a scientific problem. If the environment has recently changed, then selection may not yet have had sufficient time in which to act. In effect, selection is lagging behind the requirements of the environment. For example, the last thing we should expect is that recently domesticated animals, such as most zoo species, would be adapted to the conditions in which they have been newly placed. Their adaptations will instead reflect selection undergone in their previous habitat, in the wild. Thus, the "big cats" in zoos retain spectacular running abilities that they can't exercise. Parrots in zoos can't use their considerable brains in their depauperate cage environments. Most zoos are thus inevitably benign cruelty for most of their denizens.

SELECTION IS YOUR FAIRWEATHER FRIEND

Selection makes organisms work according to Darwin's theory. In that sense, selection is every organism's friend. Almost everything about your body that works was created by natural selection acting on randomly generated genetic variation. But selection is neither all-powerful, nor all-caring. It works effectively when the ecology, physiology, and genetics of a biological situation allow it to. But if they don't, selection may be weak, ineffective, or actively harmful. It all depends on the "weather," or rather the working conditions, that evolution faces. In fair weather, with abundant genetic variation and little in the way of side effects, selection can accomplish marvels. In bad weather, selection may accomplish little. It is hard to hold firmly in mind two seemingly contradictory facts, the power of selection and the weakness of selection, but both account for much of the complication of life, if not its actual perversity.

Darwin did not know the specific hang-ups of genetics and physiology that are now known to impede the action of natural selection. However, he had excellent intuition about them. Darwin was not one to predict the endless accumulation of perfect living things. Rather, his understanding of natural selection seems to have been that it makes animals and plants as good as they are, which is considerably short of perfection. Still, like most evolutionary biologists that have followed, he was clearly in awe of what natural selection has wrought.

4 EVOLUTION

The Tree of Life

EVOLUTION is the unfolding of heredity and selection to make the canvas of life. But long-term evolution, which an evolutionary biologist usually expects to produce most of the important events in the history of life, is one of the most difficult areas for the application of Darwinism. This arises not because biology can't explain much of evolution in principle. After all, Darwinism supplies the only well-attested model for how evolution works. The difficulties arise instead because we are usually in no position to gather enough data about the mechanisms of long-term evolution. We don't have time machines. How are we to study the events surrounding the origin of life, given that we aren't able to observe them directly? How are new species created? What are the selection mechanisms that generated such amazing characters as flight, echolocation in bats, the bizarre feathers of many tropical birds, and the sheer size of whales or redwoods? Such are the problems that arise from the vastness of evolutionary time. But these are problems that afflict all large-scale theoretical systems of life. We can at least see how Darwinism fares compared to the competition.

SYSTEMS OF LIFE

You are a member of the species *Homo sapiens*, belonging to the Family Hominidae, the Order Primata, the Class Mammalia, the Phylum Chordata, and the Kingdom Animalia. This is your taxonomic designation, which has stood for some time, since the work of the Swedish biologist Carolus Linnaeus in the eighteenth century. The basic ideas of taxonomy date back to Plato and Aristotle.[1] The most important idea in taxonomy is to classify organisms into groups based on similarity. It was natural for a Greek systematizer like Plato to organize classification hierarchically, with broader designations subdivided, and those subdivisions divided still further. What is amazing is how well such hierarchical classification works in biology. Imagine doing the same thing for the patterns of stars in constellations, the governments of present-day nations, or the music played on radio stations. In those cases, such a hierarchy seems all too often to be forced. In the case of biological species, it works well. Why should this be so?

A creationist can account for this hierarchy in terms of God's psychology. That is, the hierarchy is external to the creationist "system of life," the pattern of species which are supposed to have existed through all past ages. The actual system of life that a traditional creationist, like a biologist before Darwin, assumes, looks like this:

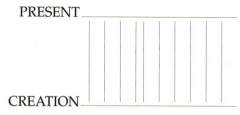

the vertical lines represent species through time, with the present at the top of the figure and the origin of the universe at the bottom. Each species is created at the origin of the earth, has been present ever since, and is still with us now. In this creationist system, the way species pass through time is not related to the hierarchical pattern of living things.

One important alternative to the creationist scheme was the system of life proposed by Lamarck, in the early nineteenth century. Lamarck is often touted as a forerunner of Darwin, as we have seen. Chapter 2 also discussed how his mechanism of evolution, the inheritance of acquired characters, differed from natural selection. However, this was not the greatest point of difference between Darwin and Lamarck. Lamarck's pattern of evolution was also radically different, as shown below. In Lamarck's scheme,

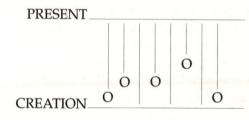

there is still an original creation of life. After this first creation, new forms of life arise spontaneously from unliving substances. (These points of spontaneous creation are indicated by O.) After origin, each species evolves into more and more complex forms independently, in the Lamarckian scheme. Simple forms of life become worms which become insects which become simple vertebrates. This pattern is repeated over and over, giving rise to all the different worms and insects and vertebrates. An insect might be part of a lineage on its way to becoming a bird, perhaps, or a turtle. New species do not usually branch off from old species, and species do not usually become extinct.

After Lamarck had had his say, English geology, particularly Charles Lyell, weighed in with its own views about the long-term story of life based on the fossil record. Lyell was the geologist from whom Charles Darwin acquired his gradualist inclinations.[2] Lyell knew enough paleontology to be aware of the fact that some fossil species disappear in the fossil record, while new ones arise. Thus, Lyell proposed that species newly originate after the first creation, as well as go extinct. However, Lyell did not propose that new species develop from antecedent ones. Nor did he propose a concrete mechanism for the origin of new species. In terms of our previous graphical conventions, we can represent Lyell's system as follows. The only addition we need is an X to represent species extinctions.

In most systems of life before Darwin's evolutionary theory, the hierarchical organization of diversity is unrelated to the pattern of species creation or extinction. The exception to this general rule is Geoffroyism (named after É. Geoffroy St.-Hilaire), a nineteenth-century variant of Lamarckism. This theory was based on the idea of multiple species descending from a common ancestral species. For this reason, the French sometimes view Darwin as a latter-day Geoffroyist. However, this school of thought never attained the influence of Darwin. In Darwin's system of life, the hierarchical organization of diversity arises because of common ancestors shared by present-day species. Species share common ancestors because the system of life is conceived as a *branching tree in which new species arise from extant species*. In Darwin's scheme of life, there is no spontaneous generation; life arises from life, except perhaps for one or a few original forms of life. When life first begins, it must come from nonliving material. Extinction is also allowed, unlike Lamarck's scheme. Darwin's tree of life is in fact more original than his theory of natural selection, though the tree actually came first.

To return to our previous graphical conventions, we have a tree pattern with the base of the tree located at the origin of life, and the tips of the branches that make it to the top of the diagram representing the species that are alive today. Again, we let O represent the origin of a new species from unliving substances. Branching of species, to create new species from old, are *not* represented with an O. Extinctions are represented with an X.

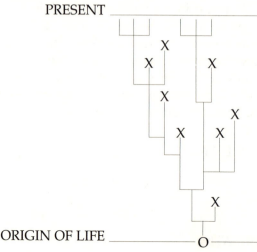

THE MEANING OF THE TREE OF LIFE

The implications of Darwin's scheme for our understanding of life are little short of staggering. First, it suggests that all of life, or at most each of a few divisions of it, must share a common physiological basis. This has been dramatically revealed since the 1950s by the universality of nucleic acids as the basis of life, DNA in particular. (A few viruses use RNA to encode their genomes.) That the genetic code relating DNA sequence to protein amino acid sequences should be virtually universal is as much a triumph for Darwin as it is for molecular biology. It was Darwin who predicted that such biochemical universality would be the case a century preceding its discovery by molecular biologists. Likewise, the ubiquity of amino acids, even the widespread occurrence of specific proteins such as histones and superoxide dismutase, suggest the essential unity of life. Again, this can be explained awkwardly in terms of the repetitiousness of a deity, or naturally in terms of an evolutionary tree relating all of life back to a common ancestor.

Second, in addition to the unity of life, the particular patterns of similarity among species can be explained in terms of common ancestry. Thus, the common features among birds, such as feathers, are to be explained in terms of the possession of feathers among the common ancestral species of all living birds. And so taxonomic hierarchy is derived from turning the tree of life upside down. Most common characters can be explained in terms of derivation from common ancestors that had similar structures, a phenomenon known as *homology*. It is in this sense that bat wings, the front hooves of horses, and the human hand are homologous: all are thought to be evolutionary modifications of the digits of the two front limbs of a mammalian ancestor common to all three groups. The differences between these three structures are explained by evolutionary biologists as resulting from adaptation, at least in part. Thus, "unity within diversity" is the product of the interweaving of similarity due to ancestry with divergence due to adaptation.

Third, evolution is necessarily continuous, much as a familial lineage is. Groups of species cannot occur twice at widely separated times. If a group is wholly wiped out, it cannot be regenerated in precisely the same form again. Evolution can only gradually work on the materials. It cannot boldly create completely new models of life. François Jacob has called evolution a "tinkering" process. Tinkers were the village and town craftsmen who would find ways to repair damaged clocks and other machines, sometimes improving them in the process. But no tinkers were in a position to invent completely new machines, like cars, submarines, or aircraft. That would be the work of the engineers who would displace the tinkers in the modern economy. Thus, until very recently, the evolutionary process has been a truly gradualist one, with an absence of miraculous novelties. At least until man appeared.

Five Points for Darwin's Tree

Darwin's tree is the most powerful idea in biology, making sense of the whole sweep of life, at least in outline. Yet it is obviously difficult to test in any simple way, unlike Mendelian genetics, for example. Nonetheless, there are empirical points which it satisfactorily explains that any other system of life would find almost impossible to explain rationally, except by saying that that was the way God did it.[3]

First, it is possible to trace the broad outlines of evolution in all forms which readily fossilize. Forms which do not fossilize, like earthworms, are difficult to study using geological formations. But lineages like those of the vertebrates, which fossilize well because of their hard skeletons, have been worked out in considerable detail. The resulting patterns are clearly like those of Darwin's tree.

Second, the farther back we go in the fossil record, which means the deeper the strata of rock, the greater the difference between fossil forms and living species, on the average. This is as we expect if evolution is a continuous historical process, accumulating small changes in an essentially irreversible way.

Third, fossils from adjacent layers of rock are more likely to resemble each other than fossils from widely separated strata. In a sense, this is just a natural corollary of the second point.

Fourth, the living forms present on one isolated land mass, like Australia, are usually closely related to the recent fossils of that region, rather than those of distant regions. This is another illustration of the continuity of Darwin's evolutionary process. It is also a blow against any rational creationist scheme, since the created species for different tropical regions ought to resemble each other, rather than the species of adjacent temperate regions, if there is any logic to God's creations. Yet the latter is typically the case. Generally, such "biogeography" is only explicable in terms of common descent—the overwhelming concentration of marsupial mammals in Australia alone makes a devastating case for this point of view. Life is produced locally, from the materials at hand, not globally by a beneficent designer.

Fifth, embryology reveals that early development involves structures which can be readily associated with common ancestors, even when these structures are not retained in the fully developed form. A dramatic illustration of this is the fact that mammalian and bird embryos develop gill arches early in embryonic development, which is only reasonable if we suppose that these groups originally had fish species for common ancestors. More generally, the embryos of quite large taxonomic groups tend to go through very similar stages early on in development. This is not to argue that development recapitulates evolution in any precise way. But the developmental process for many living things often betrays evolutionary origins quite dissimilar to those made manifest by the adult form. In other words, there appear to be "residues" in our lives, archaic bits of physiology and anatomy, that make little sense unless our ancestors were very different kinds of organisms. This is a situation that evolutionary theory can explain easily, but which nonevolutionary theories have to struggle over. Naturally, the characteristic response of most creation theorists is to ignore this point, or twist it into a criticism of evolution on the grounds that the recapitulation is not perfect. But evolutionary theory in its modern form does not require any such perfect recapitulation. For evolution, the archaic features of life merely reveal its tortuous history, like the archaic features of human language or common law.

The evidence in support of Darwin's Tree is so good, and its explanatory power so vast, that the theory of evolution by modification from common ancestors was widely accepted by biologists in Darwin's lifetime. When you hear of controversies about evolutionary biology, you can be fairly confident that they will concern the role of natural selection or variation in the rate of evolution. Darwin's Tree of Life is one of the great slabs on which modern biology is built. It isn't going to be corroded by the weak bile of Darwinism's many critics.

WHAT IS A SPECIES?

The species concept before Darwin was closely related to the concept of the Platonic organism. Species were absolutely separated from each other. Variation within species was incidental. Species were therefore to be recognized as absolute types, clearly demarcated from one another. Nothing was less appealing to biologists before Darwin than the idea that species would grade into each other, with no clear dividing line.

For Darwin, species were just names that were given to more or less differentiated forms. As such, species designations were matters of practical convenience. They introduced arbitrary verbal labels to aid research. In so doing, they somewhat masked an underlying continuity between forms.

The chief foundation for separating species for Darwin was whether or not there were intermediate "varieties," or populations, which allow some degree of exchange of hereditary material between the putative species. When such intermediates exist, then Darwin was inclined to deny species status to two distinct populations. If they do not exist, then Darwin accorded these populations species status. Out of this basic kernel, modern species concepts have arisen. But before we get to that point, note the world of difference between the Platonic species and that of Darwin. For Darwin, there is nothing fundamental about species, only a lack of exchange of hereditary material, and convenience in recognizing differentiated populations. Species have been stripped of their metaphysical status in Darwinism.

Simply speaking, an evolutionary biologist recognizes the development of a new species when the evolutionary tree branches. (This isn't the only case when new species are recognized. But it is one of them.) But what does it mean for the tree to branch? Evidently, the two species must now be evolutionarily independent of each other. This independence does not necessarily mean that the selection pressures facing the two species are different. If they live in the same region, they could well be the same. Rather, it reflects Darwin's idea that the two species are no longer in a position to exchange hereditary material with each other. In modern language, "gene flow" has been cut off.

What does the cutting off of gene flow entail? Fundamentally, it entails *reproductive isolation*, such that when individuals of two different species are brought together, they cannot mate and produce fertile progeny. In terms of how the evolutionary process works, this species concept is the natural one. After all, it focuses on the point at which species go their separate ways—when they can't mate with each other.

But paleontologists, studying fossil forms, cannot test for successful mating. Their presumptive species died millions of years ago. What are they to do in identifying species lineages? The approach of paleontologists has been to define their species post hoc. For them a species is a distinctive unit in a fossil sequence. Thus, for example, around three million years ago in Africa a hominid with large jaws and teeth arose. It was upright and many of its skeletal features were like the lineage that evolved into *Homo sapiens*, but it clearly did not do so. It remained different until it went extinct. Thus, it is accorded the status of a separate species. We can't say that individuals of this type could not have interbred with our ancestors, but their apparent morphological distinctness and separate evolutionary fate suggest that they did not.

That was a relatively hard case. On a larger scale, it is even easier to identify evolutionary independence. About sixty million years ago, a number of mammalian lineages began to differentiate into the major groups that we see today. Their origins are not known—there is a gap in the fossil record suggesting extremely rapid evolution, a problem we will discuss in more detail later. After this gap, we are confronted with more than a dozen separate evolutionary lineages. These lineages separately produced monkeys, bears, horses, whales, bats, and so on. These are easily accorded evolutionary independence, given their morphological distinctness, ecological differentiation, and disparate evolutionary fates. Such evolutionary independence, then, is the essence of the paleontological definition of species.

How New Species Are Made

Evolutionary biologists think that new species are often made in the following way.[4] An established species is, at some point in its history, split up into a number of geographically separated populations for a considerable period. These populations rarely exchange migrants, so distance acts to keep them evolutionarily independent of each other. Enough evolutionary time elapses for considerable evolutionary changes to occur in each of the separate populations, changes which are not uniform over all of these populations. Some of these populations become quite different from the rest. When the barriers to migration are relaxed or reduced, contact is reestablished between the populations. The divergent populations no longer interbreed freely with the other populations when migration occurs. They now constitute new species. For example, physical anthropologists now think that some hominids migrated from Africa to Europe six hundred thousand years ago, or more, evolving into a separate species called *Homo neanderthalis*. When *Homo sapiens* later spread out from Africa, about one hundred thousand years ago, they did not interbreed with the Neanderthals. We had become two species.

There are two variants of this kind of story, adaptive and non-adaptive. In the first, evolutionary change during geographical isolation occurs because natural selection has favored a new type of adaptation from that present in the ancestral population. After this novel adaptation is well established, reproductive isolation may occur because of a variety of reasons connected to the new adaptation. One possible factor is habitat divergence. Evolution in a new geographical region may have made a forest animal into one that prefers grassland. Even if later migration takes the grassland form back to the region where its ancestral forests were located, it may not go back to using the forest habitat. If it isn't in the forest, it won't mate with the forest animals, producing evolutionary isolation, and thus speciation. Another possibility is that the differentiated populations may produce hybrid progeny that are unfit, because they may possess neither ancestral nor new adaptations. The hybrids thus become unhappy monsters, trapped in a biological no-man's-land. In both of these scenarios, adaptation has essentially led to the evolutionary estrangement of the populations that descended from a common ancestral form. Adaptation has thereby driven the formation of new species.

There are other means by which geographically separated populations can become evolutionarily independent of each other. One of these is genetic reorganization, a slower version of the genetic revolution that DeVries found in the evening primrose, which we saw in chapter 2. Chromosomal rearrangements occur in isolated populations, and these rearrangements may fix accidentally, like the accidental fixations of words in isolated human populations. In American English, the words trunk, truck, and gasoline are used in place of the English boot, lorry, and petrol. Sometimes the accidental rearrangements of chromosomes can be so extreme that they prevent the production of viable or fertile progeny in crosses with individuals that lack these rearrangements, just as it may be nearly impossible for the middle-aged to understand the slang of teenagers. In this case, the genetic system is drifting in such a way that reproductive isolation occurs as a secondary by-product. Natural selection does not engender this type of speciation; genetic accident does.

Another type of speciation occurs when subpopulations develop that specialize on different hosts, both for feeding and for mating. This is thought to be particularly common in insects, such as fruit-eating flies. When particular fruits become the sole arenas for both feeding and mating, with an absence of migration between them, there is a kind of separation before speciation. A focus of controversy has been whether or not selection for adaptation to, say, two different fruits will lead to the splitting of the population into two groups, which then avoid mating with each other, and so leads to speciation. There are some laboratory experiments in which selection on the preference of flies for different habitats leads to reproductive isolation, providing that reproduction occurs in a habitat-specific manner. It is still uncertain how common this type of speciation is in nature.

Reproductive Barriers between Species

Speciation is the kind of process which is intrinsically difficult to observe. It is expected to take many generations. For most organisms, we are looking at hundreds if not thousands of years. Hardly the material for a Ph.D. project. Thus, our inferences about speciation tend to be indirect. Sometimes we find populations which show a high degree of differentiation from each other, but are not yet species. Hybrids can still be bred. The many populations of the fruit fly *Drosophila willistoni* in South America appear to be of this character. In other cases, distinct species are sufficiently similar genetically that we can use their chromosome differences to unravel pathways of species differentiation, one from the other. What we can't usually do is observe speciation directly.

One of the few ways in which we can obtain information about speciation is by examining the *mechanisms of reproductive isolation*. They provide evidence as to the likely factors which split species up, although this evidence is not necessarily definitive. Additional reproductive isolation mechanisms might develop after the original ones broke the species apart. However, they are the best we usually have, for now.

We begin with reproductive problems before fertilization. Examples of this are known among the *Dendrobium* orchids. These species require the right weather to flower, such as a thunderstorm on a hot day. The orchids then produce flowers which last but one day, opening at dawn and withering at nightfall. One of the species flowers eight days after a suitable storm, another nine days afterward, while a third flowers ten or eleven days later. In this way, the three species remain reproductively isolated. In most animal species, there is a characteristic male behavioral repertoire which elicits female interest. In frogs, it is calling patterns, determined by pitch, duration, and trilling. Female frogs are attracted primarily to the male call of her own species. In fruit flies, mating is based on male patterns of movement and stimulation, particularly foreleg tapping, circling, wing movements, licking the female genitalia, spreading the female's wings, and so on. When fruit fly mutants are constructed that lack part of the normal mating behavior, most females will not mate with them, particularly if they can choose normal males instead. These are just some of the ways in which plants and animals "keep to their own kind."

Even when gametes are brought into close contact, fertilization may not occur. For example, in crosses of the fruit fly *Drosophila pseudoobscura* with *Drosophila imaii*, the foreign sperm are inactivated and expelled from the vagina. Even in groups with external fertilization, such as sea urchins, it has been found that mixing the sperm and eggs of two species together still results in a preponderance of same-species fertilizations.

But once fertilization is achieved, reproductive isolation can be sustained if the offspring has negligible fitness. The sterility of hybrids has interested biologists since classical times. Aristotle was fascinated by the sterility of mules, the offspring of horse and donkey crosses. A common finding when hybrids are successfully reared to maturity is that the gonads of the hybrids do not produce viable gametes. This is true, for example, in the mammalian hybrids which are viable: horse × ass, horse × zebra, and cow × yak.

PATTERNS OF SPECIATION IN THE FOSSIL RECORD

"New species have appeared very slowly, one after another, both on the land and in the waters." So says Darwin near the beginning of chapter XI of the *Origin of Species*. Darwin imagined that there is some variation in the rate of formation of new species, but he assumed that the overall nature of the process can only be described as gradual. This is in keeping with the entire tenor of Darwin's thought, continuing on from the style of Charles Lyell, among other British scholars.

The gradualist vein of Darwin's thought was extrapolated by geologists into a highly simplified scenario of species lineages gradually diverging from each other in the fossil record. Indeed, for a time, the goal of paleontology seemed to be the delineation of more and more refined demonstrations of gradual evolution of species in the fossil record. The irony of this effort was that Darwin himself was disinclined to expect the gradual appearance of species in the fossil record: "Geology assuredly does not reveal any such finely-graduated organic chain." His reasons for thinking so were partly geological. A former geologist himself, he was well aware of processes which would cause the destruction of layers of sediment, and thus the loss of long periods from the fossil record. Even the conditions for the formation of fossils were not to be counted on. Thus, chapter X of the *Origin of Species* was entitled "On the Imperfection of the Fossil Record."

An irony of Darwin's caution is that the twentieth century has seen amazing success in the discovery of intermediate forms in the fossil record, particularly among vertebrate species. The increasing detail of the human fossil record is one example. Another is *Archaeopteryx*, the dinosaur-like fossil with teeth, claws, and feathered wings. Even the mysteries of the pre-Cambrian period are yielding to discoveries like the massive fossil formations of the Burgess Shale. Being a geologist, and a fossil-collector in his *Beagle* days, may have made Darwin overly cautious.

However, an additional point in favor of Darwin's caution concerning the fossil record was that his model of the evolutionary process itself suggests the potential for evolution to proceed in a fashion that would not lead to well-defined trends in paleontological data. One of these problems is that intermediate species may evolve specific adaptations, present in neither ancestral nor descendant forms, making a hash of attempts to find obvious intermediates between the ancestral and descendant species.

Finally, Darwin raised the entire issue of the dubiousness of fossil species, and the difficulties that that can generate for attempts to adduce particular paleontological patterns. Overall, paleontological gradualism may have been derived, in part, from Darwinism, but not with Darwin's full-bodied consent.

Darwin's rather critical and analytical perspective on the fossil record wasn't recovered until the 1970s, when Niles Eldredge and Stephen Jay Gould wrote a paper arguing that we should not expect continuous generation of species in the fossil record.[5] Rather, they argued, in considering a sequence of rock strata and their fossil forms, we should expect to see the abrupt appearance of new species, a pattern they called "punctuated equilibrium." In its original formulation, this concept was derived from the conventional speciation scenario based on geographical separation of populations. If new species are created as a result of geographical dispersal, evolutionary differentiation, and secondary contact with the original species, then the fossil record *at any one place* should show the abrupt appearance of new forms. It's like the sudden appearance of the country cousins. You're related, but you want to deny it. You can't even believe that it's true, looking at them. But that is what a long time away in a new environment has wrought. The fossil record of speciation should therefore look "punctual," and not gradual.

That much was easy enough for many evolutionary biologists to accept. Things became dicey when geologists went out and looked at worldwide distributions of related forms. Then the claims became much more intense on both sides. Some argued that global gradualism was the rule, while others found data that seemed to show punctuation throughout the entire range of some species. Stephen Jay Gould, in particular, reacted by upping the ante. His theory became that evolution was characterized by long periods of stasis interrupted by rapid evolutionary change, or punctuation. The biological basis for this punctuation was initially obscure, and Gould toyed with macro-mutationist ideas, like those discussed in chapter 2.

With time, the controversy was diffused. Population geneticists showed that even rather gradual selection within populations could produce evolutionary change that would appear virtually instantaneous on a geological time-scale, such as that defined by the fossil record. Gould backed away from some of his flirtations with non-Darwinian evolution. The main people who felt that something big had really happened were the editors who put together cover stories for popular magazines, as well as a rabble of anti-Darwinians, including creationists, who are often happy to celebrate confusion among the Darwinians.

The consensus now is pretty much where Darwin was. We expect evolution to be sedate in biological time, but its results can be fairly abrupt and disjointed in the fossil record. We are all grateful that Gould and Eldredge brought us back to a more realistic view than the gradualism of traditional paleontology.

PATTERNS OF EXTINCTION: RETAIL AND WHOLESALE

Almost all species go extinct within a few million years. This is the natural result of a living world in which species are generated according to their own particular population genetics, as opposed to some cosmic ordering principle, be it aesthetic or moral. Beautiful birds of paradise are annihilated, while cockroaches thrive. Not all species will have the attributes required to reproduce at a rate sufficient to allow their preservation in the face of the numerous sources of mortality that face every species.

In addition to this somewhat gloomy perspective on extinction, there is also the creative role that extinction plays in the evolution of life on earth. There are two aspects to this creative role. First, extinction removes those forms that are "out-of-date" physiologically and anatomically. For example, if one considers the evolution of terrestrial vertebrates, a prominent trend has been the evolution of progressively more efficient locomotion. While some of this evolution has involved the spread of more efficient adaptations within species, in some cases entire species of awkward walkers or runners have been extinguished. The evolution of the horse, in particular, is thought to have entailed many episodes of extinction or selective replacement of equine species that were less efficient at running. Still, other evolutionary biologists feel that the story of horse evolution has been dominated more by accident than adaptation.

Second, extinction may open up niches that have been occupied by "dominant" forms that had precluded the evolution of their successor species. One of the greatest stories in the fossil record is that of the extinction of the dinosaurs toward the end of the Cretaceous, a bit more than sixty million years ago. One of the consequences of this extinction event was the evolutionary diversification of mammals in the Tertiary, the geological epoch that followed the Cretaceous. Within a few dozen million years, mammalian species were to be found occupying most of the habitats that had once been occupied by dinosaur species. In this situation, extinction allowed the radiation of an entirely new group of species. However, this is not to argue either (i) that the mammals caused the extinction of the reptiles, or (ii) that the mammals are in some sense "more adapted" than the dinosaurs. Rather, the point of this example is that the extinction of the dinosaurs made possible the evolution of new types of species, whether worse or better, from a Darwinian standpoint.

The fossil record is saw-toothed, with jagged canyons of substantially fewer species separating periods in which species diversity is high. These canyons are the result of mass extinctions.[6] We have already mentioned one case which is thought to be a mass extinction, the dying out of the dinosaurs at the end of the Cretaceous. At this same time, numerous other sorts of species went extinct, from oceanic plankton to terrestrial plants. Luis and Walter Alvarez suggested that the explanation for these mass extinctions is the impact of large meteors or comets directly killing off many life-forms, with secondary climatic effects killing off still more. Among many pieces of evidence in support of this theory is the great increase in iridium, a rare element on earth that is common in meteorites, at the geological stratum defining the end of the Cretaceous. The Alvarezes proposed that the large object that hit the earth sent up ash that darkened the skies for a year or more, causing the death of most vegetation, and subsequently the demise of many species that fed on large quantities of vegetation or that in turn fed on such herbivores. It is thought that our mammalian ancestors were then small, scavenging, nocturnal animals that would be particularly well suited to surviving such a holocaust. A gigantic crater from this impact has been found at the Yucatan peninsula of Mexico. At present writing, this astronomical collision theory seems to be more or less the consensus view among evolutionary biologists concerning the extinction of the dinosaurs and their contemporaneous species, leaving aside perhaps some traditional paleontologists, who still prefer less catastrophic models.

Is It a Rubber Tree?

Mass extinctions are examples of processes that are not obviously compatible with the gradualism of Darwin's geological training. A critic might suppose, then, that Darwinism stands for nothing, and the field of evolutionary biology is bankrupt of ideas or integrity. However, this conclusion is not a legitimate one in terms of the functioning of science. Einstein and the other revolutionaries of modern physics greatly expanded the physics created by Galileo and Newton. And yet their field remains physics. It so remains because it uses the methods of analysis and experimentation that their great predecessors had pioneered. Similarly, Darwinism remains faithful to the methods of analysis of Darwin himself, including the judicious use of natural selection, descent from common ancestors, and a general inclination to gradualism. Where any of these elements no longer work, they are set aside and new ideas are employed in their stead. But the foundations are always those created by Darwin and, when appropriate, Mendel.

PART TWO

APPLICATIONS OF DARWINISM

INTRODUCTION TO PART TWO

DARWINISM is more than a great scientific tradition. It is also one of the important influences on the material lives of those who live, and have lived, in the modern world. There is nothing unusual about this. All successful science has material consequences. Only unsuccessful science is without power or application.

The most important of the applications of Darwinism is the most mundane: modern agriculture. There has long been exchange of people, ideas, and findings between agriculture and Darwinism, dating back to Darwin's own research. Since WWII, much of the great success of agriculture has been due to the application of Darwinian principles, particularly in the field of quantitative genetics.

Darwinists have recently turned their attention to issues of medicine. They have come up with some radical proposals for the reform of medical practice, particularly with respect to the treatment of infectious disease. In addition, Darwinists have discovered new tools for understanding and controlling major medical problems that were not otherwise amenable to medical research, especially the problem of aging. For medicine to be reformed on a scientific basis, it will need to add evolutionary biology to the biological disciplines that it assimilates.

It is the secret shame of Darwinism that it has played a considerable role in the founding and legitimization of a movement known as "eugenics," a poisonous phantasm of Francis Galton. Eugenics is nothing less than the attempt to apply Darwinism to the deliberate breeding of humans. As such, it has inspired the hope of utopian ideologues and the horror of those who have lived with its material consequences.

5 AGRICULTURE

Malthus Postponed

ONE OF the least known impacts of Darwinism has been its effect on agriculture. Darwin derived much of his biological knowledge from agriculture, and the flow of information back from Darwinism to agriculture has since been considerable. Some of the leading practitioners of Darwinism have worked primarily within an agricultural research context.[1] The present chapter is a brief attempt to redress the neglect of agriculture by most historians of Darwinism.

Our focus will be the single most important issue in agriculture, the extent to which it can keep up with the burgeoning human population brought about by the successes of modernity. Darwinism has played a key role in this explosive competition between human reproductive fecundity and agriculture.

MALTHUS AS CASSANDRA

Thomas Robert Malthus is known in the history of evolutionary biology primarily as the person who envisioned the scenario of geometric human population increase coming up against slower improvement in agricultural production. The outcome that he hypothesized was a catastrophic acceleration in the spread of disease, strife, and general misery. This set the stage for both Darwin and Alfred Russel Wallace, enabling them to propose that, once such a crisis of overpopulation has occurred, natural selection would then discriminate among organisms of different heritable capacities for survival.

But Malthus didn't make that intellectual leap. He was more interested in the practical question of where the burgeoning European population was going to find its next meal. He didn't see any prospect for substantially increasing agricultural productivity, so he argued for sexual self-restraint to diminish the output of new mouths. In the tradition of many modern moralists, he felt no compulsion to inhibit himself, as he had numerous children. Restraint instead was something to be practiced by others, so that Malthus's children and grandchildren could live in an orderly, well-fed world.

Cassandra was a Trojan woman who predicted the downfall of Troy at the hands of the Greeks, at least according to Homer in the *Iliad*. Apollo had given Cassandra the power of prophecy, but not the power to convince people of the truth of her prophecy. Few intellectual disorders are more common among prophets than imagining themselves unappreciated, with unique knowledge of the future but no respect in their own time. Malthus was an early example of the Cassandras of overpopulation, a forerunner to contemporary media ecologists. It must be a painful burden, to be so sure that you foresee an upcoming disaster, but to have so little impact.

THE MALTHUSIAN FAMINE DIDN'T ARRIVE

Like most Cassandras, Malthus was wrong, at least where the next two centuries were concerned.[2] Absent war, civil war, or state policy, Europe has not seen spectacular starvation since the Irish potato famine of the mid-nineteenth century. Malthus's forecast was one of general, systemic famine, not the kind of local catastrophes brought about by unfortunate combinations of plant disease and imperialist neglect, as occurred in Ireland. Indeed, Malthus was writing just before the last great European famines (1790–1850). The 150 years after 1850 were to see less inadvertent European famine than any other time since the fall of the western part of the Roman Empire, around A.D. 400. This occurred despite the fact that the geometric population increase forecast by Malthus did in fact take place during the period from 1790 to 1930 in Europe.

The historical fact is that agricultural productivity has more than kept pace with the exponential growth of the human population.

Consider grain production. The United States produced 378 million bushels of grain in 1839, at the start of the modern era, when its population was about 17 million. In 1957, the U.S. produced 3,422,000 million bushels, with a population of about 180 million. So while population size increased tenfold, agricultural production increased about one thousand times faster! Malthus was almost entirely in error. Agricultural production could be increased far faster than human population growth, in fact, and thus in principle.

Whether or not this growth in production can be sustained is certainly an arguable point. Much of the increase in agricultural productivity has depended on the depletion of three finite resources: (*i*) arable land for cultivation, (*ii*) extractable ores for the production of fertilizer, and (*iii*) fossil fuels to propel farm machinery. However, if one considers the extensive irrigation of the world's deserts with desalinized water, mining of less extractable ores for fertilizer, as well as the use of the vast, though more expensive, reserves of fossil fuel, like the Canadian tar sands, then even these three finite resources may remain abundant for some time to come, albeit at a price.

An additional constraint on agricultural productivity is the secondary damage to the environment arising from modern agriculture, from the destruction of forest to the use of toxic chemicals in fertilizers and pesticides to the greenhouse effects of diesel-powered tractors and even flatulent cows. The modern era of abundant food may be only transitory, but it has nonetheless coincided entirely with the duration of modernity. Culturally it is the only thing the young people of Western industrialized countries understand. For them, dieting is a more important issue than famine. Thus, the main experience of starvation in the industrialized West since WWII has been that of people who are too fashion conscious, augmented by the obese struggling to comply with the directives of their doctors and unfortunate young people with eating disorders. For most of the species, widespread starvation is a secondary byproduct of poverty, deliberate totalitarian policies, civil war, or other misfortune, not a natural disaster that afflicts most members of our species. As such, we have an entirely different notion of our ecological precariousness from that of a medieval peasant, craftsman, or soldier. For them starvation could still arise from crop failures and the like. Agricultural abundance is one of the essential features of the modern world.

Darwinism and the Agricultural Revolution

Consider the situation before Darwin. It was normal in pre-Christian times for Europeans to practice fertility rites to prepare the ground. But the hard truth was that these festivals of vernality did not augment harvests. In medieval Christian times, colorful rituals were replaced with prayers and payments to the Pope. It is doubtful that these worked any better. Portents of disease and inbreeding were taken to be the work of miscellaneous demons, before Christianity, or Satan, after the coming of Christianity. Thus a goat born with two heads would be taken as a sure sign that evil was about, and only (a) ritual sacrifice, (b) prolonged prayer, or (c) payment to the bishop (or perhaps a combination of these) would remedy the situation.

With the coming of Darwinism, a thoughtful pastoralist would instead conclude that there might be a breeding problem and consider obtaining a new he-goat for his herd to provide some relief from inbreeding. This cultural transformation occurred because Darwin's research provided a materialistic basis for biology. Though Darwin did have a credulous belief in such phantasms as the inheritance of acquired characters, he placed them entirely within a materialistic account of heredity. Darwin did not seek divine intervention as a source of the fittedness of animals and plants to their circumstances. Thus, Darwin played an important, indirect, "cultural" role in freeing the practice of agriculture from superstition.

In this general way, Darwinism helped make it possible to develop scientific agriculture. Part of that agriculture would come from the physical sciences, that part dealing with soil chemistry, erosion, the design of farming machinery, and so on. Another part would come from improved basic biology, such as techniques of sexual fertilization, both plant and animal. But part of the new agriculture would involve the specific application of Darwinian thinking, and ideas first developed fully by Darwin himself. It is to these ideas that we now turn.

DARWINISM AND APPLIED BREEDING

It is a popular myth, at least among scientists, that scientific theories determine technological development. Thus scientists often imagine that the technological superiority of Europe from 1700 to 1900 was an outgrowth of the research of Galileo, Newton, Laplace, and other physicists.

That this is a myth is revealed by the many distinctive European inventions that came before the scientific theories that supposedly inspired them. Pendulum clocks, accurate maps, and telescopes all came before the hypothesis of the Newtonian universe. Indeed, it could be argued that the first good telescopes were the primary impetus for the creation of Newtonian physics. Using such telescopes, Europeans first saw through the classical theory that "the heavens" were entirely different from "the earth." The early telescopes showed that the moon had what were then thought to be mountains like those on Earth. (It would turn out that the geology of the Moon wasn't like that of the Earth either, but that discovery came much later.) The moon wasn't a perfect sphere, contrary to Greek astronomy. Motion in the heavens might be the same as terrestrial motion, contrary to Aristotelian physics. European physics could abandon its classical heritage.

Technology, particularly new instrumentation, can determine the development of science. Seeing new things, measuring new variables, may inspire new scientific theories. In such cases, science can be seen as *post hoc* scrambling for an interpretation after technology reveals a new feature of the world. In other cases, admittedly, scientists have created new technologies, a preeminent example being Michael Faraday's invention of the electrical motor. The point here is only that the flow of information is not always one-way, from science to technology.

There are few cases in the history of basic science for which this general principle of driving or motivating technology is truer than evolutionary biology. But in this case the technology involved is not that of gleaming machinery. Instead, it is the technology of plant and animal breeding. Humans have kept domesticated plants and animals for millennia. Perhaps none have been domesticated longer than the dog (*Canis familiaris*), for which domestication began about one hundred thousand years ago. And dogs show many overt signs of having been domesticated. Despite all being derived from one wolf species, dogs now show spectacular heterogeneity among dozens of breeds. There are large, powerful breeds, like the mastiff, St. Bernard, and Newfoundland; there are small dogs, like the Chihuahua, Pekinese, and Yorkshire terrier; and there are fast dogs, like the greyhound and wolfhound. There are dogs that are all black, dogs that are all white, and every kind of color and pattern in between. Under domestication, dog breeds have diverged spectacularly. Indeed, some of these artificially created breeds could not survive long in the wild and thus are dependent on man. Similar, if less pronounced, results have been obtained for cats, horses, cattle, sheep, wheat, potato, rice, and so on.

As Darwin himself realized, some of this differentiation has been the result of deliberate selection and some has been a result of what he called "unconscious selection." Deliberate selection has occurred in many breeds of animal, particularly "working breeds." Dogs that are bred to help with the hunt have been selected for very specific traits, sometimes to track down prey (as with hounds), sometimes to root out prey (as with some terriers), and sometimes to retrieve wounded or dead prey (as with retrievers). Dogs that fail to perform according to expectations are not used for breeding purposes. Indeed, they may be destroyed, like the errant sheep-dog of Thomas Hardy's *Far From the Madding Crowd*, which pushed an entire flock of sheep over a cliff, ruining his master. But other breeds probably came to their particular features from inadvertent selection combined with inbreeding. Thus, Darwin comments, we have an English "Spanish pointer" that was derived from a Spanish breed but no longer has any resemblance to any extant Spanish dog.[3] New breeds can come into being without any intent on the part of any breeder.

But the important point here is that Darwin's own ideas on natural selection were strongly shaped by animal breeding. The first chapter of the *Origin of Species* is virtually an introduction to animal breeding, particularly the breeding of pigeons. In his intellectual development of natural selection, Darwin is clearly appealing by analogy from human breeding of animals to "breeding" in nature. In effect, the achievements of human breeders were generalized to the workings of nature.

As we saw in chapter 3, a number of experimental biologists tried to test Darwin's theory of evolution by natural selection. The failed experiments of scientists like Johannsen, Jennings, and Pearl, among others, showed that experimental breeding was not an easy thing to do. You had to have genetic variation to start with. Then you needed characters that were reasonable, if not easy, to measure. And you needed some kind of experimental controls to help you monitor other experimental factors, like inbreeding, deteriorating lab environments, and so on. Well-executed experiments didn't start until the 1910s, particularly with the work of William Ernest Castle using hooded rats, as we have already seen.

But these experiments were basically prehistory in the development of scientific breeding. A key ingredient was missing before 1918: the genetical theory of selection. That theory began with the work of R. A. Fisher, mentioned in chapter 2. Fisher's first publication of a population genetic analysis of quantitative variation appeared in 1918. Between 1918 and 1952, when Sewall Wright published a magisterial synthesis,[4] theoreticians created an explicit body of theory for what came to be known as "quantitative genetics." In everyday language, quantitative genetics was the application of Darwinian and Mendelian ideas to the variation of characters like height, weight, survival, fertility, and so on. The key to understanding quantitative genetics is that it considers variation in easily measurable characters, like height, instead of variation of specific genes. Quantitative genetics assumes that genes play some role in determining quantitative characters. But the genes are moved "offstage," like puppeteers. All we see are the puppets, the visible body parts. For agricultural applications, this was essential, because there was little likelihood that animal and plant breeders were going to be able to identify all the genes affecting the characters that they wanted to shape using selection. And like the successes of dog and cattle breeders from prehistory, it was clear that progress could be made without identifying the genes.

Since the 1950s, almost all Western breeding for agricultural purposes has taken place within the context of quantitative genetics. My first teacher of quantitative genetics, Eli Scheinberg, was fond of saying that he didn't teach quantitative genetics—he taught corn and hog breeding. Even to this day, works entitled *Evolution and Animal Breeding* are published,[5] works in which Darwinism, Mendelism, and agriculture are all mixed together.

If one visits a local fast food restaurant, as we all must from time to time, and bites into one of their Mega Meals with hamburger, fries, and shake, very little that one eats has been left unshaped by selection. The cow that provided the beef has been bred for "marbling" of fat with muscle, so that the beef will be tender when cooked. The potatoes in the French fries have been bred for resistance to blight. The vegetable oil in which the French fries were cooked came from corn that has been selected to be oily. The milk in the milkshake came from cows that give vastly more milk per day than in 1900. Even the ketchup that you add to the hamburger or fries came from heavily bred tomatoes.

Agricultural breeding has been one of modern man's defining achievements, in fact if not in glamorous press coverage. Agricultural productivity has been increased manyfold by the application of selective breeding. A large part of our present population would not be here, were it not for the great increases in yield produced by the application of scientific breeding techniques. In other words, Darwin and Mendel are the people ultimately responsible for refuting Malthus.

How Scientific Breeding Works

Plant and animal breeding is so easy, but so powerful, that it is not surprising that humans were breeders long before they had any conception of what they were doing. Nonetheless, it is important to have a good algorithm for selection in order that it may operate efficiently.

The start of any selection program is genetic variation. If there is no genetic variation for a character, then there is no point selecting on it. On the other hand, numerous studies of selection and genetic variation have revealed that almost all measurable characters exhibit some genetic variation: height, weight, fat content, muscle mass, etc. Thus, it will rarely be the case that the minimal requirements for selection will be absent. Note that genetic variation is a *necessary* requirement, not a *sufficient* one.

Given a genetically varying breeding stock, the first step of a good selection design is to rear a generation together from the same point in time under the same conditions. Such a group of organisms is generally called a cohort. Uniform rearing practices are very important. If we are breeding sixty cows on each of three farms, then there is the possibility that one of the farms provides better feed than the others. The cows from that farm then grow faster. If we choose the largest cows in our selection experiment, then the cows from one farm will dominate among the selected for reasons that have nothing to do with their genetic value. Rather, their conditions of rearing will have been the determining factor. So here we will assume that you didn't make this elementary mistake, and all your cows have been reared uniformly.

At some predetermined age, you measure all your cows for the same character in the same way. Let's say you are measuring them all for body weight. Then they should all be measured on the same scale at the same age at the same time of day after the same meal. Suppose that their average weight is 300 kg, at the chosen age. Having made your measurement, you choose the top 10 percent of the cows.' Say these selected cows have an average weight of 350 kg. The "selection differential," the difference between the selected group and the rest of the population, then is 50 kg.

Your unselected cows are sold to Mega Meal, Inc. for hamburgers, and you breed from the selected top 10 percent of cows. Furthermore, let us suppose that you did the same with your bulls, and their selection differential is also 50 kg, the weight of chosen bulls minus the weight of all bulls. If you breed only from the chosen cattle, and rear their offspring in the next generation, you expect an improvement in their weight in the next generation. Say you get an average in the next generation of 330 kg. Your "response to selection" is 30 kg.

Note that the response to selection is less than the selection differential. In part, this reflects the influence of the environment. Some of your selected cows and bulls were big because of accidental qualities of their environments, not their genotype. To a first approximation, quantitative genetics tells us that, so long as the pattern of genetic variation does not change, the same magnitude of selection should produce the same increment in the character each and every generation. Therefore, you should be able to continue selecting each generation and get roughly the same response to selection, something around 30 kg. In this way, animal and plant breeders extrapolate from the first one or two generations of selection to future generations of selection. Such extrapolation has probably played a large role in the willingness of businesses to fund animal and plant breeding.

You Can't Always Get What You Want

When was the last time you took a plastic-wrapped tomato out of its package and bit into it? You may have experienced the bland taste and indifferent texture that now characterize tomatoes. Tomatoes used to be larger, redder, and tastier. They also spoiled faster and were more expensive to pick. Something about the breeding of a tomato that was easier to grow and pick has made them inferior in taste.

This example illustrates one of the most important principles of breeding. You can't select for only one thing at a time. Inevitably, other characters respond to selection too. Genes that code for characters like body size may also code for other attributes as well. For example, genes that foster growth hormone production may give larger cows. But they may also produce more docile cows. (This example is hypothetical.) Most hormone secretion patterns are interrelated; if one hormone level is altered, several others may be as well. In this case, size and docility increase together because of an inherent interconnection, physiologically. There may be no way by which the selection procedure could be altered so as to break this correlation. It is a by-product of the fact that functional characters, like size or activity, are dynamically interconnected, not severable.

When animal and plant breeders produce plants or livestock that are cheaper for mass production, they may also have unintentionally bred for food that tastes like cardboard. This isn't because they want us to be unable to differentiate the food from its packaging. It arises instead because they can't get exactly what they want from their breeding without introducing other changes that are not appealing. Therefore breeders assume that, during the course of selection, the selected character is unlikely to be the only character responding to selection. These other responses to selection may cause increases in undesirable characters, gradually turning many of our foodstuffs into imitations of tasteless fiber.

Darwinian Limits to Modern Agriculture

The fact that Malthus has been beaten by modern agriculture should not be taken as grounds for unlimited optimism about our future supply of food and other products of agriculture. Indeed, there are general Darwinian grounds for expecting that we will run up against biological limits that will cause a dramatic erosion in the ability of humans to continue increasing in numbers without grave misfortune. Specifically, our continued freedom from famine has depended on the use of scientific breeding to produce better stocks of animals and crops. There are reasons to fear that this cannot continue.

The most elementary problem limiting progress in breeding programs is the exhaustion of genetic variability during selection. Selection requires abundant heritable variation to progress. But when all beneficial genes are fixed, there is no more variation to use in selection. This problem of using up variation is a common finding in long-term selection experiments. It shows up as plateaus in selection response after many generations of selection. With plateaus, there is no continued response to selection. No more progress is made. This is such a common result that it is virtually the normal expectation among breeders. Some laboratory experiments that have used exceptionally large population sizes have obtained long-sustained responses to selection. Such results show that natural populations are probably not likely to be subject to the exhaustion of genetic variation. But agricultural breeding populations are often small, especially if they are farm animals that have already been selected for a long time. There may be millions of steers, but there may be only a few thousand bulls, in any particular breed of cattle. It is the number of bulls that limits the size of the breeding population. Therefore, the occurrence of plateaus in selection for improved grain yields, cattle growth, or dairy production is to be expected on the most elementary quantitative-genetic reasoning.

Darwinians realized a long time ago that the response to selection could be continued if new genetic variation is introduced into populations that have become stuck on plateaus. Indeed, animal breeders have been using this practice for centuries. Nothing is more common among breeders than the use of semen from a new bull to improve herds that have failed to make progress with selection. Inbred lines of dogs, likewise, are often rescued from health problems by crossing in new sires, sometimes leading to the creation of a new, more robust breed. This practice lies at the foundation of an interesting theory of evolution propounded by the aforementioned American evolutionist, Sewall Wright.[6] Wright felt that there would be definite limits to the progress that selection could make in individual lines or breeds. His conclusion was that they would tend to get "stuck" in evolutionary dead ends. Rescue, he felt, would come from more successful populations sending out migrants. The introduction of new genetic variation in the populations receiving such migrants would then evolutionarily pull the inferior stocks toward the more successful condition. Like many of Darwin's ideas that came from the practices of animal breeding, particularly of pigeons, Wright built his theory on established practices of cattle breeding. It was yet another case of "technology" leading the way for science.

Hybrid plants are among the most important crops. The significance of hybrids is that they often remedy the deficiencies of the two parent stocks. It is a common enough practice to take a hardy, disease-resistant, but small-fruited plant and cross it with a domesticated strain with abundant fruit, but little disease resistance. Frequently, though not always, the hybrid will combine the merits of both parents, having disease resistance together with abundant fruit. However, these benefits often do not survive a generation or two of continued cultivation of the hybrid. This is to be expected, because crosses of hybrids with hybrids produce some of the original parental types. For that reason, hybrid grain must be regenerated for each growing season, using crosses of the original parent lines. The production of such hybrid grain is one of the biggest and most successful parts of agricultural breeding programs.

Access to genetic variation is the ultimate limiting factor determining the progress of scientific breeding in agriculture. Therefore, the total number of different varieties or strains of a species in turn limits the future of agricultural productivity. Many of these varieties come from nature or non-Western societies. For example, the potato is of South American origin, where dozens of varieties are to be found. Western agriculture employs only a small fraction of the varieties found in the world. But the spread of scientifically selected Western stocks is threatening to undermine continued cultivation of the more diverse South American potato stocks. And this story is only generic.

Western agriculture and urbanization are also destroying the habitats of many wild varieties of grain and other plants. We are causing the extinction of a number of ungulate species, some of which are our best prospects for replacing the domestic cow, should it be lost to us as a result of disease. Any Darwinian would rather have the possibility of greater genetic diversity in a species of economic importance. Such genetic diversity allows continued progress during selection. Genetic diversity is also insurance against catastrophic elimination of livestock or crops as a result of disease. The Irish potato famine killed one million people in part because the potato variety being used by the Irish as their staple food simply wasn't genetically diverse enough to survive a single disease. We can only hope that Malthus won't have his point made by the narrow stupidity of an agricultural economy that is heedless of its Darwinian foundations.

It is not only the genetic foundations of crops and livestock that lie at the root of the vulnerability of modern agriculture. Since many of our agricultural species are, or will become, lacking in the genetic variation with which to respond to environmental change, there is the possibility that famine might result from environmental deterioration itself. The problem is exacerbated by the degree to which modern agricultural breeds are dependent on ideal environments. These include good housing or planting conditions. Such conditions can be undermined by global warming or the depletion of the ozone layer, both of which may result in greater stress for domestic animals and plants alike. These organisms are also often pathetically dependent on the provision of nutrients, antibiotics, fertilizers, pesticides, and so on. The problem is not just one of survival. The domesticated stocks may indeed survive. But if they have been selected for high performance under ideal conditions, their genetic superiority under those conditions may not translate to ecological conditions that are substantially different. The scenario being sketched here is the loss of the unnatural conditions that have progressively characterized domestication in the industrial countries, and indeed have become established through much of the world. If these conditions are lost because of ecological dislocation, or pervasive economic collapse, then well-established, high-performing breeds may give very poor performance indeed. The result, again, could be widespread famine.

On the other hand, we have the following view:[7]

> The main argument in Malthus's book has, since that time, been disproved by events. "There has only been one man too many on this earth," said Proudhon, "and that man was Malthus."

6 MEDICINE

Dying of Ignorance

DARWINIANS often encounter M.D.s who have biological opinions that are discordant with basic evolutionary research. Among the most important of these is the idea of "old age" being an inevitable wearing-out of body parts, essentially unchangeable. Another is the notion that all the symptoms associated with disease are necessarily pathological. Still another is the idea that all striking deviations from the norm are disease-states. And these examples could be multiplied.

The prevalence of these half-baked and ill-formed ideas has recently provoked George C. Williams and Randolph M. Nesse, among others, to reformulate medicine on Darwinian foundations, an effort called "Darwinian medicine."[1] At the moment, this is more a theoretical paradigm than an alternative program for medical practice. But the hope is that someday medical practitioners will have their work as enriched by Darwinism as the practice of agriculture has been. The primary beneficiaries should be the patients.

There are many elements to Darwinian medicine, just as there are to medicine itself. One of the central ideas is the importance of evolutionary history. Another is that fertility is a major issue for medicine, not a minor inconvenience or luxury. Perhaps the arena where evolutionary biology can make the greatest intellectual contribution is infectious disease, where our uninformed intuition may be a poor guide to the complexities of the relationship between humans and their pathogens. Aging is an area where medical science has been essentially impotent, which makes the considerable success of the evolutionary biology of aging heartening for the goal of someday postponing human aging substantially. The far frontiers of medical science (cloning, genetic engineering, and the like) are coming closer every day. An evolutionary perspective suggests that the radical redesign of humans may not be as easy as some have supposed.

INSECTS DON'T GET CARDIOVASCULAR DISEASE

The central idea of Darwinian medicine is that the evolutionary history of an organism dictates its medical problems. There is nothing inevitable about many of the medical disorders that afflict humans. Or at least there is nothing inherent in the basic organization of life about these particular disorders. It is evolutionary history that is the foundation of disease, where that evolutionary history has given us both medical problems and medical health. In the evolution of particular organisms, genetic variation and natural selection act with great power and finesse to produce marvels of biological adaptation. The gills of the fish, the lungs of the whale, the wings of the bat, the rump of the Mandrill baboon, these are among the achievements of evolution. But these achievements each give rise to particular vulnerabilities, in two ways.

First, some of these adaptations contain the seeds of internal disarray and cumulative malfunction. Examples of these are cardiovascular disease and cancer. Insects are effectively free of both of these problems. Their circulatory system is open. The blood basically sloshes around in an open reservoir. So deposition of plaque on the walls of arteries is not a problem. Insects also have very little cell division in the adult body, most of it being confined to the gonads. They occasionally get tumors, but these are always benign in genetically normal insects. (Insects that suffer from malignant tumors can be produced by mutation.) This is not, however, a Darwinian criticism of mammals having closed circulatory systems and adult cell proliferation. There are many benefits that accrue from these adaptations, efficient respiration and a proliferative immune system being two. But these biological adaptations are associated with distinctive pathologies, which are now the leading causes of death in industrialized countries, cardiovascular disease and cancer, respectively.

Second, some of these adaptations provide opportunities for infection by pathogens. Two prime examples are the respiratory tract of terrestrial vertebrates and the reproductive tracts of animals with internal fertilization. Such structures invite infection for two basic reasons. The first is that gas exchange and gamete transfer both require appropriately moist tissues. Otherwise, the cell surfaces won't exchange oxygen and gametes will die from desiccation. But moist conditions that are beneficial for the basic functions of lungs and vaginas are also excellent for the invasion of viruses and bacteria. The second reason respiratory and reproductive tracts invite infection is that they are conduits for pathogens because of their relationship to the surrounding environment. This is obviously true in the case of internal fertilization, which provides the opportunity for gonorrhea, syphilis, HIV, and other diseases to flourish. But respiratory tracts are another excellent venue for diseases, since air must be inhaled actively. Pathogens that can establish themselves near noses and mouths have every prospect of being sucked into the body. Yet again, these adaptations may be essential to survival and reproduction for terrestrial mammals with internal fertilization, while at the same time fostering disorders that induce death and suffering.

In short, by each of these two mechanisms, inherent design problems and exposure to infection, evolution by natural selection dooms organisms to their particular cycles of disease and death. Some types of organisms get one type of disease, other types get other diseases. The evolutionary details matter, for in our beginnings are our endings.

A natural reaction to these extreme contrasts would be to dismiss their practical relevance. After all, there is no prospect of our developing an open circulatory system for patients with cardiovascular disease, so why should these evolutionary critiques be of any interest to the practicing physician?

There is nothing more basic than choking. We have all had the experience of eating a sandwich while talking, when suddenly things go wrong. We can't breathe, we feel panic, and we try to cough. Or rather, our bodies try to cough, because it isn't a decision that we are making reflectively. Suddenly, you're fighting for your life from the simple act of eating a sandwich. In a chilling analysis, Nesse and Williams take us through the basic biology of choking. About one person out of every one hundred thousand chokes to death each year. This arises because our ancestors were simple aquatic animals. In those ancestors, input involved water only. From water, we obtained both oxygen and small food particles.

Water was expelled from gills and shunted down the gut. Over time, this situation changed for some species thanks to evolution, particularly land-dwelling vertebrates like mammals, birds, and reptiles. We developed mouths, noses, lungs, and stomachs, and four passageways associated with each of them. The problem is that all these passageways meet in the pharynx, located between the rear of the mouth and the opening of the esophagus. While there are valves and other devices in your throat that try to direct traffic, sometimes they fail. Food or water go down your windpipe, blocking the passageway to the lungs. Your body tries to open up the windpipe. If it succeeds, you live. If it doesn't, you die. Death by evolutionary history.

Interestingly, human babies, which otherwise seem so vulnerable, can breathe and swallow at the same time. This must reflect their unusual evolutionary history as appendages to the breasts of biped females that, formerly at least, nursed for long periods. Horses can do this trick, too.

Reproduction Is the Meaning of Life

The brown marsupial "mouse" of the genus *Antechinus* isn't really a mouse. It is less related to true mice than normal mice are to us. It lives in the woodlands of eastern Australia, a seasonal environment. The life cycle of the male is usually played out within the span of a single year. Like many organisms, there is a single mating season in the year. In the weeks before this mating season, sex hormone levels, such as those of testosterone, begin to increase dramatically in males. Other hormone levels also rise, particularly corticosterones, hormones that are associated with stress in mammals. During the mating period itself, males become highly aggressive toward each other. Fighting can be extremely violent. Copulation likewise can be remarkably intense. Marsupial mice can sustain coitus for up to twelve hours, a large amount of time by any measure, but particularly so relative to the length of the lifespan and the size of the animal.[2]

Beginning just before the mating period, and then continuing afterward, males show many dramatic pathologies. One of the most insidious of these is that they cease to groom their fur, a symptom common to moribund mammals. Despite ingesting food, they lose weight. The testes and spleen shrivel. Roundworms, protozoa, and bacteria proliferate in the male body, suggesting pervasive immunosuppression. Bleeding ulcers give rise to anemia. Death occurs within two to three weeks of the mating season. Females, by contrast, do not exhibit these changes. Their fur remains in good condition and their hormone levels remain normal. In many cases, females can survive for another year, sometimes breeding again. Why do males exhibit this spectacular crash after reproduction? Why are females so different? What does this phenomenon say about the workings of evolution in determining pathology?

These questions can be answered at two levels, physiological and evolutionary. In terms of pathophysiology, the key is the effect of castration. If males are castrated before the mating season, they do not show high levels of hormones, weight loss, or organ degeneration. Even their fur remains sleek. In field tests, castrated males survive longer. Castrated males are less aggressive toward each other. In the laboratory, they can survive up to three times their normal lifespan. The conclusion is obvious: testes kill marsupial mice. Get rid of them, and their health improves. Females don't have testes, and that is why they do so much better.

In terms of evolutionary biology, male marsupial mice exhibit a "big bang" pattern of reproduction. They have evolved to put all of their resources into a single consuming act of reproduction. Most notably, from an evolutionary biology standpoint, this is the "healthy" pattern for these males. A castrated male is an evolutionary nullity, in the absence of paternal care for offspring—and such care is absent in most male mammals. Thus the longer-lived, sleek, apparently happier, castrated male is a disaster in Darwinian terms, because it can't reproduce; it can't transmit its genes to the next generation. There is thus a perfect antagonism between a medical definition of health and a Darwinian definition of health. Since the medical profession is almost entirely oriented around the maintenance of life, the little detail of castration for the marsupial mouse is greatly outweighed by the spectacular survival benefits. Since Darwinians view function entirely in terms of net reproduction, a castrated male is seen as a complete failure, absent some compensating contribution to close relatives, as in the kin selection situations described in chapter 3.

An interesting point is the perspective of human culture generally. If one reads ancient texts about the maintenance of health, fertility is an extremely prominent concern. The number of myths about lost male potency, to give just one example, is staggering.[3] "Phallic" deities are among the most common mentioned in the *Larousse Encyclopedia of Mythology*. The Greeks had the classic Priapus, a deity whose name is incorporated in English in priapism, meaning persistent erection of the penis. Indian cosmology has two major potency gods, Puchan and Prajapati. It is no surprise to discover that such phallic deities are often located at the top of theological pyramids, one early example being Amon from ancient Egypt. Along with his strictly phallic role, Amon represented the forces of generation and reproduction as a god of fertility. Female fertility too is a source of great human concern. Indeed, to imply that a female was barren was once one of the most devastating of insults. After phallic deities, there are also numerous deities for successful childbirth throughout the world's cultures. Classical Greek deities like Aphrodite have become specifically associated with swooning infatuation, thanks to bowdlerizing Christian culture. But the original deity was considered the goddess of love and sex in all their forms, from romantic passion to erotic degradation to fertility. The broader culture of every society, however suppressed by the puritanical, always seems to contain nooks and crannies in which the weeds of sexuality and anxiety about fertility grow vigorously. People in their everyday lives often deal with their health as if fertility is a central concern.

This is not to advocate the maximization of human fertility at every opportunity, only to point out the interesting relationship between three different systems of value: the medical, the Darwinian, and that of everyday people. The Western medical perspective may be out of step with the needs and desires of everyday people, regardless of its ethical correctness by its own standards.

HAZARDS OF BEING MALE

There are any number of health problems associated with being a man, particularly a greater risk of early heart failure. In the modern world, there is a common inferiority of male longevity compared with female longevity.[4] There are some exceptions to this pattern, which may be revealing. In Syria, Pakistan, Bangladesh, India, and Iran, men and women have life expectancies at birth that are within a year of each other, the male living slightly longer in the last three countries. In Nepal, men have life expectancies that are 2.8 years greater, 50.9 years for men versus 48.1 years for women. It is interesting to note that most of the countries that have rough parity of life expectancy between the genders are not entirely modern, typically lacking extensive industry, public health facilities, and so on. In countries where all those things are in place, and overall life expectancy is high, women consistently live longer. Notable examples include Finland and France, where the life expectancy differences between men and women are 8.4 and 8.2 years, respectively. In the United States and Canada, the differences are 6.9 and 7.1 years, respectively. Thus in "advanced" countries, being male is a major threat to your survival.

If one were a muckraking "health" journalist, and this were any other group than males, this situation would be a source of great outrage. But since some people in the media seem more moved by the plight of women, as illustrated by the contrast in coverage between the comparably deadly breast and prostate cancers, this massive health problem goes largely unnoticed. It is a widespread penalty of modernity, which does not differentiate between cultures or diets as different as those of Japan, Canada, and Finland. Therefore, one can't accuse men eating steaks of gender-destroying behavior. You can eat sushi and die sooner too, if you're male. Thus, the usual gimmicks of the food and health media won't plausibly protect men from the hazards of being male.

The solution to the empirical puzzle of early male death is to be found in evolutionary biology. If one knows a lot about zoology, then one is not the least surprised by different survival patterns in the two genders. Indeed, gender is one of the most important determinants of survival patterns. There is nothing mysterious about this situation. Male and female bodies are different launching platforms for genes. With the differences in production of gametes, in parental care, and in competition for mates between the two sexes, different reproductive patterns will evolve in the two genders. Males may seek numerous mates, compete violently with other males, and contribute little parental care. Females may mate with just one male and invest heavily in parental care. Or the genders may exactly reverse these roles, as they do in sandpiper species in which the small, male birds are left to tend the nest. Some fish species have male parental care too. In any particular animal, the genders may be very similar or very different in their reproductive behavior. But, as we have seen in the case of the marsupial mouse, reproductive adaptations can have devastating effects on survival.

Thus, there is some reason to suppose that human male reproductive strategies may curtail survival, relative to female reproductive strategies, at least under the conditions prevailing in the modern world. The obvious test of this hypothesis is castration.[5] Understandably, this is a difficult experiment for recruiting volunteers. However, back in the good old days of Dickensian medicine, medical doctors thought little of castrating male patients they considered mentally retarded or insane. From such "experiments," it has been found that the mortality rate of hospitalized eunuchs is decreased compared to matched controls that have also been hospitalized, giving a greater lifespan. However, under the conditions of long-term medical care, both normal and castrated males died much sooner than males that live outside of hospitals. This isn't great evidence, but it does suggest that the hormonal determination of "maleness" by the testes gives rise to some or all of the male survival deficit. It should also be noted that there is a wide range of data in other organisms, like marsupial mice and Pacific salmon, which show an increase in male survival with castration. It's unlikely to be hamburgers that are the problem.

So, in a sense, the medical villain is the testicle. But at the same time, the testicle is the essence of manhood. Castrate a male mammal before puberty and he becomes a very different animal, less aggressive, usually with relatively more fat and less muscle. The eunuchs of the Pope's choirs or the Sultan's harems were very different from intact men, with different voices and different patterns of growth. Often they were taller, because testosterone inhibits the growth of long bones like the femur, which largely determine height. Thus, the genitals are the biological center of a man.

There are many puzzles associated with the male genitalia. One is the long route taken by sperm from the testicle to the penis. This trip is taken using two vasa deferentia, one from each testicle. The vasa deferentia pass up the body, over the pubic bone, around the ureters, through the prostate gland, and then finally into the urethra near the base of the penis. The urethra is then the launching tube for the concupiscent sperm. The irony of this trip is that the vasa deferentia pass right by the urethra early in their subdermal

journey, but proceed onward for some distance before finally join-
ing the urethra. The normal male's method of preparing sperm for
ejaculation is thus about as efficient as the office of a governmental
bureau, in which every task has to have as many complications as
possible.

The reason for this awkward anatomy is evolutionary history, in
particular the intertwined evolution of the male reproductive tract
and the male kidney. Indeed, at one point in our ancestry, the testis
sent sperm through the primitive fish kidney and thence on to the
nephric duct and through it to the outside world. Female mammals
have not had such intertwined anatomy, so they have no tangling
of oviducts and ureters. But the male vertebrate has long had this
interaction, so human males have their sperm plumbing all en-
meshed with their urinary plumbing.

Another puzzle is the location of the testicle. The testicle is the
only unique organ which is located entirely outside the skull or the
main body cavity. Of course our muscles, blood, mammaries, and
sensory organs are located outside the protective recesses of the
body. But it is less consequential to lose a finger or an ear than it is
to lose organs like the heart, lungs, brain, and so on. Thus, it is easy
to see why the human body is constructed in such a way as to protect these vital organs. In women, the ovaries are located well inside
the body, largely safe from outside threats. But in the man, the entire Darwinian meaning of his existence dangles in a vulnerable little sack called the scrotum.

Present thinking on this subject revolves around temperature. It
has long been known that tight underwear, hot baths, and fever can
reduce male fertility. The testicle also has a heat-exchanging set of
blood vessels which indicate that its circulatory system has been
selected to cool the testicle down relative to the rest of the body, by
about seven degrees Fahrenheit. Analogously, birds are known to
reduce their body temperature during the period of spermatogene-
sis. Testicles are not unique to humans, being the normal anatomi-
cal structure encasing the mammalian testes and their associated
plumbing.

Birds and mammals share a common major adaptation, homeothermy. Both groups actively maintain elevated body temperatures independently of the ambient temperature. There are many
physiological and ecological advantages associated with a stable,
high body temperature. Lizards, which lack homeothermy, have

great difficulty being active at night or early in the morning. Their metabolism remains sluggish until they are fully warmed up by their environment. Rodents, on the other hand, can easily function at night, possibly eating the eggs of the lethargic lizards at such times. Even more dramatically, homeothermy allows birds and mammals to live in the polar regions, enjoying great advantages from their maintenance of speed and strength under such cold conditions. But high temperatures are bad for DNA replication, and thus bad for making sperm, which are disproportionately rich in DNA compared to all other human cells. Men make about one hundred million sperm a day, each sperm packed full of DNA. Women do not make new eggs as adults. Thus, the descended testicle may be a price that men pay to be homeotherms.

OUR MUTATIONAL LOAD: NATURAL SELECTION AS AN UNDERACHIEVER

A common misinterpretation of Darwinism is that it is a guarantee that good things will happen to living things. After all, it might be said, shouldn't the increase in fitness that Darwinism predicts make all biological attributes approach perfection? Not at all.

One of the most profound factors limiting the perfection of organisms is recurrent deleterious mutations. Many of the birth defects that we see are due to these mutations, though by no means all. In addition, some disorders exhibited only later in life are caused by mutation, examples being Huntington's disease, Werner's Syndrome, hairy pinna, and so on.[6] There are a number of distinct ways in which mutation rains on our parade. The most extreme are the mutations that entirely prevent survival into adulthood. Many of these cause miscarriages and stillbirths, which are traumatic for the parents, but do not see much light of day. Still others give rise to a viable infant who only later becomes dramatically afflicted.

One of the most terrifying genetic disorders is childhood progeria. This is thought to be caused by a dominant mutation. It doesn't run in families, apparently because it renders each and every carrier sterile. Thus every victim has suffered a new mutation. There is little outward manifestation of progeria during infancy. Beginning in childhood, around age five or six, victims of this disorder start to develop overt symptoms. Their rate of growth slows, they lose their hair, their skin begins to look old, they suffer cardiovascular disease, etc. These children go from youth to the semblance of extreme old age in a few years. Tragically, they remain normal intellectually. They have no protective dementia or amnesia to obscure the horror of what is happening to them. Very few live to be teenagers, and none survive into adulthood. Reproduction is not a possibility.

This is only one of many afflictions that can strike children as a result of new mutations that only require one copy to be expressed. There are extensive manuals listing hundreds of other disorders of this kind.[6] The obvious question is, why doesn't natural selection prevent these mutations from occurring? There are two answers that can be given. The first is that natural selection is not a perfect screening system. A dominant mutation that gives zero Darwinian fitness, like progeria, will leave no descendants in the next generation. But new cases of these mutations will be produced at a rate equal to *twice* their mutation rate, because we have two copies of each of the genes that aren't located on our sex chromosomes. With a dominant mutation, only a single mutation is required to produce the disease, and we have two targets for these mutations. Natural selection can only screen these mutations *after* they have occurred. Thus, we will see these diseases recur. With less deleterious mutations, it takes longer for natural selection to eliminate each disease gene from the population, so they will tend to accumulate to even higher frequencies. For those genes that decrease Darwinian fitness by only a few percent, and are virtually anonymous, the frequency of the deleterious gene can rise to hundreds of times the mutation rate, in the range of one in ten thousand. Many medical problems have their origins in genetic problems of this kind.

A second answer to the underachieving of natural selection addresses the deeper problem of why natural selection doesn't eliminate mutation in the first place. Here the answer is in part that selection does indeed reduce mutation rates. Organisms have numerous molecular devices for repairing DNA which greatly reduce the mutation rate from the level it would otherwise have. If that is the case, it might be wondered, why hasn't that repair machinery evolved to eliminate mutation completely? At once, it should be admitted that nothing in the material world can be made entirely glitch-free. Mutations will always occur. However, there are reasons for thinking that we still don't do that well. These reasons are that there will be mutations in the DNA coding for the DNA repair machinery, and these mutations may cause only slight imperfections in the machinery, such that deleterious mutations then occur only rarely. But such slight weaknesses in the machinery will be only weakly selected against, if the total load of mutations does not rise too high. So, to paraphrase Woody Allen in *Love and Death*, it's not that natural selection doesn't care, it's just an underachiever.

The Oedipal Syndrome

One of the most profound features of life is that incest is a very bad thing indeed. Otherwise, Hamlet could have married his mother and saved us from five acts of great Shakespeare. Freud would have had much less to complain about, though he doubtless might have found something. And then think of poor Oedipus. He should have been able to go to the Olympic games, having happily remained with Jocasta.

So why is incest so bad anyway? The products of incestuous matings are much more likely to be mentally retarded, sterile, or stillborn. Children of first-cousin marriages have twice the death rate of children from marriages of unrelated individuals. This in turn occurs because of the increased homozygosity of very rare, recessive, deleterious genes.

Where do such genes come from? Again, they come from mutations. But not mutations like the dominant mutations just discussed. When genes are deleterious but fully recessive, in the sense defined in chapter 2, they have bad effects only when they are combined in pairs. If such a gene is kept rare by natural selection, then combinations of two such genes of the same type will be even rarer. For example, if a bad recessive gene only occurs in one in a million copies of that gene, an individual carrying two copies of such a gene will have a frequency of around one in a million million, when mating is random. Highly deleterious recessive genes of low frequency have little effect on a population that mates at random.

But mating isn't always random. Consider the situation of a courting couple. If they are both overtly normal, they may or may not be carriers for particular genetic diseases. But their chances of afflicted offspring will be on the order of one in ten thousand to one in one hundred million, over all such diseases. However, many genetic disorders are at much higher frequencies in particular areas than these numbers suggest. The reason for this is incest and other types of consanguineous matings.

Consider the mating of two siblings, when the mother is a carrier of a mutation for a recessive genetic disorder.[7] With a rare genetic disease, it is unlikely that the father is also a carrier for the same disease, and we can neglect that possibility with little error in our calculations. Each child has a chance of one-half of receiving the mutant allele from the mother. The final probability of an incestuously produced grandchild having the disease is one-sixteenth, because the chance of both siblings carrying the disease gene is one-quarter, while the probability of one of their offspring then having the disease is also one-quarter. These two events are independent of each other, so the odds multiply, giving one-sixteenth. By comparison with the mating of unrelated individuals, in which the risk is one in ten thousand or less, this is an enormous increase in genetic risk. It is further extended by the fact that *both* mother and father will be carriers of mutations for distinct genetic disorders. It has been estimated that each person carries about six such mutations. If we assume that the genetics of each mutation is independent of every other, then unrelated couples might face odds of genetically afflicted children of about one in a thousand or ten thousand, while mated siblings would have about a 50 percent chance. Incest within immediate families is little short of disastrous at the genetic level.

Diseases for the Greater Good

Proponents of Darwinian medicine are fond of pointing out that some genetic diseases can be quite severe, and yet quite common. The reason for this is that a gene that produces a genetic disorder when homozygous, and thus present in two copies, may be beneficial when present in only one copy. Indeed, individuals with two different versions of a gene may be superior to normal individuals that entirely lack the allele for the genetic disorder.

This general principle is illustrated well by the gene for sickle-cell anemia, discussed in chapter 3. Individuals with two copies of the gene for sickling have severe medical problems arising from the frequent deformation of red blood cells. These medical problems include internal bleeding, shortness of breath, and chronic pain. But sickled cells are also protective against malaria. Where malaria is common, as in Africa, the sickle-cell allele rises to high frequencies, in the tens of percent. This is a situation where natural selection actively fostered the spread of a genetic disease. Why does it do this? Understanding this apparently perverse result of natural selection requires understanding what natural selection tends to do. Natural selection does not act to improve the health of each and every organism. Instead, it acts—at best—to increase the *average* fitness of the members of the population. This means, firstly, that some individuals may be ill-served. It also means, secondly, that the focus is fitness, not the health of the body as a whole. One can calculate the health benefits of being physiologically preadolescent. Twelve-year-old humans in modern welfare states have some of the highest survival rates of any animal ever known. If we were all to forego adolescence and stay twelve forever, we would live for about 1,200 years, eventually dying of accident or infection, in most cases. But natural selection is about reproduction, and so we are hormonally programmed to proceed on to adolescence. Our death rates then immediately rise. Yet that is fine with natural selection, which trades some risk of mortality for its most important payoff, reproduction of your genes. In this sense, we are like marsupial mice.

CURES AND DARWINIAN HOLIDAYS

The fictional Victor Frankenstein's response to the inevitability of death was to use medical science. So he built his monster and reanimated its body parts. He wanted to conquer death. In its initial impulse, this aspect of heroic medicine is understandable. Of particular interest here is when medicine acts to change the selective outcome, saving the genetically disabled from death, possibly enabling them to reproduce. In so doing, medicine distorts evolution. Not when it merely alleviates the suffering of the doomed or the viable, but when it changes net reproductive rates for particular genotypes. That is, medicine can be a major player in the field of human evolution.

One of the best examples of this in the context of deleterious mutations is the medical treatment of phenylketonuria (PKU). This is a common genetic disorder in which the metabolism of a particular amino acid, phenylalanine, is disrupted in those who have two copies of the PKU gene. Untreated, children with PKU become mentally retarded. Medical doctors can now routinely screen for this genetic disorder, and such screening is required in most American states. The defective gene is fairly common. About one in a hundred Americans carries one copy of the gene, so that around one out of every forty thousand infants has PKU. We are now able to treat the condition quite successfully by giving these infants diets free of phenylalanine. Because doctors are doing this, we are largely eliminating selection against the allele. This will result in an increased frequency of the disease gene, and thus in individuals that would suffer from the disease, in the absence of treatment.

LIFESTYLES OF THE PESTILENTIAL AND FAMOUS

Before the twentieth century, people used to die primarily of influenza, tuberculosis, and pneumonia, among other infections. Heart disease and cancer were minor causes of death before 1900, though well-known medically. In London in 1651, tuberculosis caused 20 percent of all deaths. The great plagues of Europe from time to time took even more, up to two-thirds of some cities and provinces. The Black Death of the fourteenth century was probably the most devastating of the epidemics of premodern Europe. Its total mortality varied between one-eighth and two-thirds of the various local populations. England, for example, is thought to have lost about half its population. In aggregate, perhaps twenty-five million lost their lives to the Black Death, at a time when the total population of Europe would have been under one hundred million. As recently as 1918, a worldwide pandemic of influenza A took millions of lives, perhaps as many as the number lost in combat during WWI. In our time, the world's most dramatic epidemic is HIV, which has killed several million people already. However, under conditions of malnutrition and little medical care, millions of children still die each year from diarrhea or complications of colds and other minor infections. Pathogens will always be with us.

A great illusion of modern times is that diseases like heart disease and cancer are the major killers. For most of mankind throughout most of history, this has hardly been the case. Indeed, disorders like heart disease and cancer are primarily age-related problems that are best considered in the context of aging. (This we will cover shortly.) Leaving aside the temporary lull in the onslaught of contagious disease that the developed world enjoyed from 1960 to 1980, much of the history of medicine has been dominated by contagious disease. That is why Louis Pasteur is one of the most important beneficiaries of mankind. His work in the nineteenth century firmly established a scientific approach to the medical conquest of disease. Indeed, it is probably his work on pathogens and antiseptic conditions that has made a visit to the doctor a net positive for the patient's life expectancy. Certainly this wasn't the case before Pasteur. This very successful line of work was not particularly Darwinian, except perhaps in its preclusion of spontaneous generation in the genesis of disease. Pasteur was very concerned to show that, once all possible sources of germs had been eliminated, new infection required exposure to a pathogen. This was essentially a "life comes from life" doctrine, very much in keeping with Darwin's general stricture against continuing *de novo* origins of life-forms. But this was not a major factor in the development of modern medicine.

To a Darwinian, one of the most interesting things about contagious diseases is their strategy.[8] What has to be borne in mind is that pathogens are evolving organisms, subject to natural selection to maximize their successful propagation. There is a wide range of alternative strategies for pathogens to adopt. They may be specific to the human species, like syphilis and smallpox, or they may infect a variety of mammals, like influenza. Infections may be acute, like the cold or Ebola viruses, or they may be persistent, like tuberculosis or HIV. They may be deadly, like HIV or the Ebola virus, or they may be relatively benign, like the cold viruses. They may be highly contagious, like the influenza viruses, or relatively poor at transmitting themselves to new hosts, like tuberculosis. Each of these alternatives determines the medical possibilities for treatment. At the same time, these alternative strategies also define alternative Darwinian possibilities for the pathogen and, to some extent, ourselves. That is, we are not just suffering rounds of epidemiology with our pathogens. They are also important parts of our evolutionary environment, as we are of theirs.

Particular pathogens constitute unique and complex evolutionary arenas. However, there are some principles that are general to all these relationships. The first is that pathogens do not evolve to be cruel to humans or animals. Rather, their evolutionary goal is to reproduce. Pathogens that quickly kill their hosts before they can infect a new host with their offspring are going to be selected against. For that reason, there is little likelihood that humans will ever face a naturally occurring pathogen that kills in a few hours, because such diseases would face an acute problem of infecting a new host. Rather, most deadly diseases, like HIV or bubonic plague, should progress slowly enough to infect a new host. The artificial pathogens of germ warfare, of course, need not follow these rules.

The second general principle is that pathogens face a basic trade-off between specializing on one host-type versus infecting a variety of potential hosts. The advantage of the latter strategy, illustrated by influenza viruses, is that if the pathogen is eliminated from one host species, it can maintain reserve numbers on another host. This strategy does face some problems, however. Among these is the problem of getting from one host species to another. In the human case, we make this a lot easier by keeping farm and other domestic animals in close proximity to us. Another problem is that the specific biochemical machinery that is best for infecting one host species may not be best for another host species. Complete specialization, on the other hand, allows very close matching of the pathogen to the host evolutionarily, such that the pathogen may achieve very high transmission rates, from host to host. However, any particular host may, at some point, evolve complete resistance to the pathogen, causing the latter's extinction. Thanks to medical intervention, particularly vaccines, we have been able to wipe out smallpox. And since there are no other possible hosts for the disease, it is now extinct. However, it should be noted that there is considerable evidence that the end of smallpox was hastened by human evolution. European populations around 1500 had endemic smallpox, but mostly survived. When New World populations were exposed to the disease, they contracted it and died at horrifyingly high rates. Thus, European populations had apparently evolved some degree of resistance to the disease before vaccines were developed.

A third basic principle for the coevolution of diseases and hosts is the importance of the particular mechanisms of transmission from host to host. In some animals, these mechanisms can be positively bizarre, including cases where pathogens modify the behavior of their hosts to guide them toward infecting others. The deranged aggression of animal rabies victims may be an example of this. Many pathogens have seized upon the one context where human behavior can be counted on to supply frequent opportunities for infecting a new host: sex. As previously pointed out, human sex is a wonderful thing for a pathogen, because it involves internal fertilization. This requires an intimate copulatory event in which at least two bodily fluids will be produced, seminal and vaginal. Other possibilities for transmission include bleeding from open genital sores or other lesions. A veritable feast of opportunities is thereby provided to viruses, bacteria, and protozoa to get from one human to another during coitus. And of course they take advantage of these opportunities, from gonorrhea, chlamydia, and syphilis to herpes, HIV, and various nonspecific bacteria and fungi. Less "pestilential and famous" pathogens make do with other mechanisms, particularly transmission from mouth to mouth and hand to hand. Indeed, the handshake is perhaps the single most barbarous Western custom. From a human disease standpoint, it is third only to the kiss and to coitus as an opportunity for contagion. The bow is much preferable.

A fourth principle is that microbial pathogens can evolve within their hosts during a single infection, staying ahead of natural host defenses, sometimes even staying ahead of medical treatment, such as antibiotics. This is a major factor in the great tenacity of HIV infection. The human body may be able to develop antibodies against the initial infectious agent, but new variants may be produced by the pathogen that require different antibodies. This is a kind of arms race which a pathogen, like HIV, may often win.

THE BODY'S FORTIFICATIONS

At the core of human resistance to disease is the vertebrate immune system. Vertebrates are able to recognize foreign molecules as such, and then launch a counterattack on those molecules or on cells that have them. This feat is almost miraculous, because each pathogen differs from normal human cells in a unique way. How does our body recognize these foreign invaders as such, and then respond?

The answer is that the vertebrate immune system runs on Darwinism.[9] It randomly generates recognition proteins, and then keeps producing those that fit the foreign, or "antigen," molecule. This system works because the cells of the immune system undergo "somatic" mutation, mutation that does not affect the DNA of the gametes. This somatic mutation of the immune system involves both direct DNA sequence changes and the shuffling of modules of coding DNA, like cards in a deck. Up to one hundred million different candidate protein molecules can be produced by a human immune system, though only a few of them will fit the antigen well enough for further use. The specific cell lineages that correctly produce "antibodies" that match the antigens are then amplified in number by an internal proliferative control system, so that plenty of antibodies are produced. The primary cell types involved in this type of antibody production are the "B cells."

The second layer of defense is provided by the "killer T cells." Again, these cells generate antibodies by somatic mutation and differential proliferation of antigen-recognizing cells. T cells go further than B cells, in that they kill any cells that have the antigenic molecule. Those are likely to be cells harboring pathogenic virus or bacteria. The central importance of T cells is revealed by the plight of HIV patients that have progressed to AIDS, acquired immune deficiency syndrome. They have essentially lost their T cells, and thus are vulnerable to a wide range of infections that our immune systems normally dispatch with ease. "Boy in the bubble" children have congenital immune deficiency, which requires that they be stringently protected from pathogens. In either case, survival prospects are grim indeed. The immune system is vital to a normal human existence.

An interesting feature of immune systems is their "memory." Once a particular pathogen has generated an immune response, it will encounter a much more vigorous response on subsequent infection. A key qualification here is that this is true for only a particular pathogen strain. A pathogen of the same general type, say another influenza A virus, may evolve quickly to become a "new" pathogen as far as our immune system is concerned. Indeed, many of the most important human pathogens, from HIV to malaria to influenza, have adaptations that help them to evade the memory of our immune systems. In a sense, there is an evolutionary arms race between our immune system and pathogen genetic systems, an arms race fully analogous to the encryption-decryption "races" that go on between intelligence agencies.

The vertebrate immune system is one of the greatest achievements of evolution, but it is not entirely without costs. Obviously, the maintenance of the many cells involved in the immune system is one such cost. Most insects are probably too small to sustain such a complex pathogen-fighting system, though they do have some immune responses. An additional set of costs arise from misfires of the immune system. One of these is the range of problems that we know as allergies. Allergies produce cold-like symptoms or asthma in response to numerous airborne substances: pollen, spores, dander, and microscopic fecal matter. Allergic responses also arise from skin contact and ingestion, where the allergen may be wool, shrimp, or peanut. Some of these allergic responses can be fatal, particularly if the tissues of the throat or lungs swell persistently, causing suffocation. At present it is unclear why allergies occur at all, at each of the three levels of selection, genetics, and etiology. However, it is certain that they are produced by inappropriate responses of the immune system, and they are usually at least a nuisance.

Another inappropriate response of the immune system is autoimmune disease. Autoimmune diseases run the gamut from rheumatoid arthritis, potentially disabling, to lupus erythematosus, often fatal. In myasthenia gravis, for example, patients make antibodies against muscle receptor proteins, interfering with normal nerve transmission to muscle. One effect of this pathology is that muscles are not normally stimulated, causing atrophy. First this affects the "voluntary" musculature, but later the muscles that sustain breathing are affected, sometimes resulting in death. All these disorders have in common the failure of the immune system to differentiate self from not-self correctly.

These are just some of the problems with having an immune system. However, the calamity of AIDS and its congenital analogs shows convincingly that human life without the vertebrate immune system is insupportable. Like many products of evolution, the immune system isn't perfect. It can even be a disastrous affliction. But for almost all of us, it is a great boon.

The immune system is not the only type of defense that the human body mounts against contagious disease. The body also does a number of other things to kill, impede, or expel invading pathogens. Interestingly, some aspects of medical practice may interfere with these efforts to stave off disease. The simplest cases are those where the body is attempting to expel a pathogen. There are six basic tricks that we use: crying, sneezing, coughing, vomiting, copious urination, and diarrhea. And in response to these behaviors, we obtain prescription and nonprescription drugs that suppress them. These suppressive attempts of ours may, however, be wrong-headed. For example, experimental infections with bacteria followed by drug-suppression of diarrhea result in a doubling of the period of acute infection. More straightforwardly, vomiting up toxic food is a basic adaptive mechanism that must have saved many lives; not everything that ends up in our bodies should stay there.

Fever is probably the most misunderstood of all our normal defense mechanisms. Few medical interventions are more common than the reduction of body temperature to normal levels. However, this may be helping pathogens more than ourselves, at least with moderate fever. There is a wide range of evidence showing that fever reduces infection in vertebrates. Julius Wagner-Jauregg won a Noble Prize for showing that malarial fever increased the recovery rate from syphilis thirtyfold. Reducing fever seems to prolong illness or exacerbate its symptoms, from nasal stuffiness to septic shock. Greater caution in the normal treatment of fever is worth considering. Of course, extreme body temperatures must be treated, and occasional use of medication may be needed for unusual circumstances when short-term relief is required. Otherwise, "don't take two aspirin and see me in the morning" may be better medical advice.

AGING: THE GHOST OF CHRISTMAS FUTURE

There are several basic things to understand about aging. The first is that not all living things age. Bacteria probably never do. Some grasses and shrubs, like juniper, do not apparently age. Even some sea anemones, multicellular animals, don't age. They just go on growing and dividing, so long as conditions are good. This isn't the immortality of a Greek god; animals that don't age can still die. They just lack an acceleration in the rate of dying due to pervasive physiological deterioration during adulthood.

The second point derives from the first. There is no absolute biochemical, molecular, or cellular necessity about aging. Otherwise, the potential immortality of sea anemones would be impossible.

The fact that we are all products of mammalian cells that have been propagated for tens of millions of years at once explodes numerous non-Darwinian theories of aging based on some general quirk of biochemistry, like DNA repair, oxidation, and the like.

The third point is that the organisms that can live forever all have some type of vegetative reproduction, particularly fissile reproduction, where the entire organism splits into two symmetrical products. Bacteria and fissile sea anemones split in two, for example. Some potentially immortal grasses and shrubs, on the other hand, have less defined vegetative reproduction. By contrast, organisms like insects and mammals, which apparently always undergo aging, have no form of vegetative reproduction. There is something about having to have sex to reproduce that leads to aging. Identifying that something has been a project of evolutionary theorists through much of the twentieth century.

The best way to understand the evolution of aging is to consider two very different genetic diseases. One of these has already been introduced, progeria. This disorder strikes children, and it entirely prevents reproduction. Within one generation, the gene is eliminated from the population. Then there is Huntington's disease, also lethal, also caused by a single dominant gene.[10] This gene is quite common in some areas. It hasn't been fully contained because its bad effects aren't expressed until after its carriers have had at least some opportunity to reproduce. Indeed, the reproduction of carriers can be considerable. This gene is prevalent because the force of natural selection against a lethal gene is weaker when the deadly effects are only expressed later. To consider an extreme point, if a human gene kills its carriers whenever they reach ninety years of age, there would be no selection against such a gene, because there is no more reproduction after that age.

Natural selection is thus expected to do a good job of making humans healthy at twelve, but a poor job of making us healthy at eighty-two. It carefully screens genes with early bad effects, while it lets mutations and other genetic problems with bad effects only at later ages accumulate. And so we age, because natural selection is the ultimate source of our health. There is evidently nothing else in this world that takes care of giving us bodies that work. This is not a demonstration that we should be perfect when we are twelve. There are still problems like contagious disease, occasional mutations, and so on. But on average we are enormously decrepit at eighty-two compared to twelve. The obvious question is, given that evolutionary biologists have a useful scientific theory of aging, what can be done about it medically?

The answer goes as follows. If aging is as simple in origin as a lack of natural selection at later ages, then increasing the force of natural selection at later ages should postpone aging. That is what the evolutionary theory implies. And that is what is observed, primarily in experiments with the lab fruit fly, *Drosophila*. The force of natural selection acting on survival can be increased by delaying the first age of reproduction in an entire population. If you are kept waiting to reproduce, and only later are given the opportunity, you have to be in adequate health for the task. Most important, you must have survived to the later age. Thus, forcing a population to reproduce only at later ages forces natural selection to work harder at all ages before the delayed start of reproduction. This will make evolution pay attention to genetic aging problems. In effect, the underachiever is being forced to work harder. If this regime of delayed first reproduction is sustained for the entire population over a dozen or more generations, a genetically variable population can evolve increased lifespan and improved later health.

This artificial evolution of postponed aging has been shown multiple times in several laboratories.[11] Lifespan has been doubled in both mean and maximum. Later fertility has been enormously increased. Flies that have been bred for longer life are also generally more robust physiologically. They can fly for longer periods. They can resist acute stress better. Together, these results show that we can make aging evolve in the laboratory. Most important, we can use laboratory evolution to postpone aging. There aren't any fixed limits to the lifespan of particular species. Aging is evolutionarily tunable, and thus it must be genetically or biochemically malleable as well.

The next target for intervention, in principle, could be humans. The problem with this next step is that it would be both very slow and very unethical to shape human evolution in such a way as to postpone human aging selectively. One requirement would be to make the reproduction of teenagers or twenty-somethings illegal. As this is not feasible, ways have to be found to intervene in human aging without intervening in human evolution. One solution is to do enough basic evolutionary and genetic research that we can imitate what evolution would do, if it was set the task of evolutionarily postponing human aging. The key to such research will be the creation of longer-lived substitute organisms using evolutionary and genetic techniques, organisms that could reveal how to postpone aging in humans. Mammals are the best organisms to use, because they are more likely to be similar to humans in their genetics of aging. But it is also possible that there are aging mechanisms general enough that discoveries made in simple invertebrates, like insects, could give important insights. This whole area of research may be the single most promising application of evolutionary biology to problems of medicine.

From Professor Hyde to Dr. Jekyll

Darwinians have long neglected the practice of medicine. This may have occurred particularly because of the greater status of medicine compared with biology during most of the history of Darwinism. Nonetheless, present-day life sciences have seen a coming together of medicine and biology generally. Such interchange, if not fusion, has been based primarily on work in molecular and cell biology. This may have happened because of the widespread presumption that molecular biology and biochemistry provide necessary and sufficient foundations for all of biology.

But Darwinism is based explicitly on the notion that molecular biology and biochemistry do not provide a sufficient foundation for biology, or the life sciences generally. Thus, it has only been a question of time as to when Darwinists would move forward and claim the relevance of their field to a medicine that is increasingly scientific in its foundations. In some cases, this has enabled Darwinists to supply interesting analyses for well-understood medical problems. One example of this would be the population genetic analysis of human genetic diseases. In other cases, Darwinists have been able to solve problems that molecular and cell biology have almost entirely failed to resolve, aging being perhaps the preeminent example of this. However these specific cases work out, medicine without some Darwinian foundations would, without doubt, be medicine that has hobbled itself needlessly.

7 EUGENICS

Promethean Darwinism

ACCORDING to Greek mythology, Prometheus was the most brilliant of the Titans, so Zeus gave him the task of creating humanity. Because the Titans had been subjugated by the new Olympian Gods, led by Zeus, Prometheus sought to humble these new gods by giving men greater powers than Zeus had intended. So it was that, in the course of equipping the human species for its life, Prometheus went up to heaven to get fire from the sun and bestowed the gift of fire upon mankind. This generous gift of fire incurred the wrath of Zeus, who had Prometheus chained to a rock on Mt. Caucasus while an eagle fed on his liver. Each night the liver regrew, and each day it was devoured again. But Prometheus remained unbowed, expressing his hatred and resentment of Zeus with undiminished fervor. After thirty thousand years, Hercules slayed the eagle and freed the errant Titan with the permission of Zeus. Prometheus then joined the pantheon of Olympic gods. Thus it came to pass that the Athenians revered Prometheus as the benefactor of mankind and the father of all arts and sciences. They raised an altar to him in the garden of Plato's Academy.[1]

But of course none of Greek mythology is true. We weren't created by Greek, Etruscan, or other gods. Instead, we were created by long sustained patterns of natural selection acting on genetic variation present in hominid populations. Realizing this basic, not to say fundamental, truth about our origins is one of the most important intellectual events experienced by moderns.

Our Darwinian provenance does not, unfortunately, forestall those who would be latter-day Platonists from dreaming dreams of preeminence. If evolution is indeed the key force that makes us, that defines us, or that limits us, then the Promethean mission can be recast as the effort to control the direction of evolution. For many modern intellectuals and ideologues, the apprehension of the Darwinian message was only one quick step before the development of Promethean *hubris,* and the desire to seize control of the evolution of the human species. This program of deliberately directed human evolution has a name: eugenics. (The word is derived from a Greek term that means "of good stock.") The eugenics episode was the saddest and ugliest in the history of Darwinism. It revealed a demonic aspect to the Darwinian Spectre.

Victorian Racial Hygiene

Eugenics, as originally conceived, was the application of "good breeding practices" for the improvement of the human species. This was a Victorian, particularly Victorian English, idea. Plato, in his *Republic,* proposed the deliberate breeding of superior individuals with others of their kind, as well as the prevention of breeding by inferiors. But his scheme was concerned with the maintenance of good qualities, not their improvement. In 1865, Francis Galton, Darwin's cousin and the slayer of pangenesis, was the first to suggest prominently the value of controlled human breeding, where he saw such controlled breeding as producing the improvement of the species. Galton first used the word eugenics in 1883. In those days, before Mendel had been rediscovered, it was difficult to be very concrete about the types of breeding practices that would in fact improve the human stock. The main idea then seems to have been the ill-defined fostering of reproduction among the talented and virtuous, along with the imprisonment or sterilization of the habitually criminal or otherwise antisocial.

But the idea picked up momentum anyway. One factor was the commonplace obsession with race, class, and standing in nineteenth-century Europe. Another was Galton's propagandizing, in popular articles and in his book *Hereditary Genius* of 1869.[2] His arguments were examples of virtually circular reasoning. Galton began with the assumption of "natural ability," attempted to characterize it by the use of biographical and other reference works, and then concluded that natural ability was a genuine phenomenon after all. Most important, families that had high reputation or standing were more likely to have offspring who were themselves eminent in some way, in Galton's analysis, because of inheritance.

For example, he wrote in 1865, "Out of every hundred sons of men distinguished in the open professions, no less than eight are found to have rivaled their fathers in eminence." The neglect of familial environment, resources, and encouragement in this assertion is breathtaking. Perhaps a mitigation is that, at the time, the genetic model for inheritance was not widely known. Indeed, in Victorian biology, the influence of environment on inheritance was widely assumed, even by Darwin, as we saw in chapter 2. Thus Galton's naive reasoning through the minefields of inheritance and environment affecting human attainments was more characteristic of his time than idiosyncratic. In any case, his conclusion was that "natural ability" and "genius" were hereditary. The study of heredity would become Galton's lifelong obsession, because he couldn't imagine better means of improving the lot, character, or abilities of mankind.

The charming thing about Victorian eugenicists is that they hardly had a clue about how to realize their ambitions. They resembled young children, staring at their reflection in unplugged televisions and imagining that they are seeing broadcast programs. So of great importance for the Victorian eugenicists were "great families," and these typically were overachieving, well-educated, meritorious people rather like the eugenicists themselves. Ironically, one of the important pedigrees for this theory was that of the Wedgwood-Darwin-Galton families, in which there were two first-cousin marriages, a genetic fiasco. But the eugenicists chose to ignore the problem of inbreeding in this case. Victorian eugenicists took their prejudices about class and "breeding," in that very special English sense, projected them onto Darwinism, and proceeded to admire the prospect of a eugenics that would enshrine them and their values long into the evolutionary future.

In those early days, most forms of eugenics were "positive," focusing on the encouragement of the better to breed more. Thus, Galton and others proposed breeding competitions in which families were to be judged for their fitness (literally), and given prizes to encourage their further breeding. One can still find pictures of the families that have won first prize in these contests. Were they oblivious to the moral equivalence between them and prize bulls or sows? One can only conclude that those were different times. Another difference about those times was the casual proposal of "negative" forms of eugenics, in which the unfit, infirm, and criminal were to be perpetually imprisoned, sterilized, or killed to prevent their further contribution to the inheritance of the species. This was to become the most massively implemented part of the eugenics program, albeit in a way that the Victorians could not have remotely conceived.

THE COMING OF MENDELIAN EUGENICS

With the rediscovery of Mendel, eugenics could be based firmly on *both* genetic and Darwinian ideas. Indeed, with the rediscovery of Mendel, it became possible to make evolutionary biology as mathematically and statistically cogent as any branch of the physical sciences. The fumbling of Darwin and Galton with the problem of heredity, described in chapter 2, was to be replaced by the technically precise work of the early geneticists, at first particularly the botanists. Later, Thomas Hunt Morgan was given a Nobel Prize for his lab's work on fruit fly genetics, work in which genes were firmly located on chromosomes. All of these advances made plausible the idea of creating a powerful eugenics, a eugenics that could steer the future self-creation of man. It was to be a future with our species benignly guided by a select group of geneticists and evolutionists, the New Prometheans.

Unfortunately for this scenario, the coming of genetics led to much sharper calculations of the prospects for success from eugenic methods. In 1917, R. C. Punnett, Balfour Professor of Genetics at the University of Cambridge, studied the difficulty of eliminating feeblemindedness, based on an unrealistic model in which it is determined by a single gene.[3] (The lack of realism favored eugenics.) He began by assuming that the frequency of hereditary feeblemindedness is 3 per 1,000, which was then a widely accepted number. If the genetic assumption is that these individuals all have two copies of a feeblemindedness gene, then about 10 percent of the population would carry a single copy of the gene. If all feebleminded individuals were killed or sterilized, a "negative" eugenic procedure, how long would it take to reduce the frequency of feeblemindedness in the population as a whole? The answer turns out to be chilling. It would take about eight thousand years to reduce the incidence of feeblemindedness to 1 in 100,000. It would then take an *additional* twenty thousand years to reduce the incidence to 1 in

1,000,000, on Punnett's hypothesis of the genetic basis for the character. There is little in this analysis that would survive close scrutiny as to matters of fact these days. Many loci influence mental abilities, and they are variable in their mode of inheritance. However, many of these genetic complexities would only make it harder for eugenic methods to give any improvement. Punnett's calculations, therefore, stand as an early example of theoretical population genetics, strongly suggesting that the eugenics program for human improvement might rest on extremely shaky foundations. Though eugenic methods might work in principle, the amount of time required for them to do so was very considerable, essentially prohibitive. But, as with other nineteenth-century ideologies, eugenics would take some time to die.

American Eugenics

In the United States during the first half of the twentieth century, eugenics reached a high-point in its influence among both scientists and government administrators in the English-speaking world.[4] A moderate number of laws and bureaucratic directives took on a eugenic slant, if not an explicitly eugenic rationale. The most important figure in this development was Charles Davenport, Director of one of the first institutes for the study of genetics and evolution, at Cold Spring Harbor, Long Island, New York. Davenport was more of a hard-working empire-builder than a brilliant thinker. He wasn't the charismatic figure that so many in the British eugenics movement were, from Galton to Pearson to Fisher. Instead of persuading the people, he studied them, accumulating massive pedigrees of human subjects.

Davenport had many characteristic prejudices of the American scientist of his time. He was very middle-class in his values, and censorious about sexual immorality and petty crime. He was also patronizingly racist. Americans of African origin were assumed to be inferior, as were a vast range of other ethnic groups, such as Southern Europeans. Indeed, following the practice of European culture generally, Davenport tended to confuse the concepts of genotype, race, and ethnic origin, as if there were any such thing as, say, "the Italian," conceived of as a biological type. In practical terms, Davenport believed in encouraging good, middle-class, northern Europeans to breed. He favored "positive" eugenics to that extent. But his great obsession was preserving such breeding stock from contamination with "retrograde elements," such as the criminal or the so-called feebleminded. For them imprisonment and forced sterilization were the treatments of choice, "negative" eugenics.

In a typically ironic turn of history, just as the scientific basis for eugenics was becoming less credible, the work of Davenport and his ilk began to produce significant popularity for eugenics. On the "positive" side, American families would compete with each other for eugenic certification at state fairs, like prize heifers. On the "negative" side, American legislatures and courts began to implement the ideas of Davenport and his fellow eugenicists, ideas that led to sterilization and deportation. In the United States, fortunately, wholesale infant euthanasia has been unknown. More common has been sterilization of the "feebleminded," more than sixty thousand such operations by 1961. And laws allowing these sterilizations remain in force in a number of states as of the 1990s. However, the main focus of eugenics in the U.S. has been on immigration, particularly that of "inferior races." Calvin Coolidge, for example, pontificated that, "America must be kept American. Biological laws show that . . . Nordics deteriorate when mixed with other races." Keeping the blood pure was once again a popular cause, now sanctioned by the approval of science, the specious science of the eugenicists. In 1924, Congress overwhelmingly passed, and President Coolidge quickly signed, an immigration act that would effectively shut down immigration from the many countries and ethnic groups that the Protestant, "Nordic," American eugenicists regarded as undesirable. The politics of this event were complex, and the trade unions were supporters of some importance, but the patina of scientific justification was supplied by the eugenics movement.

RACE AND RACISM

But race, one of the key concepts of the eugenics movement, has been progressively dismantled by evolutionary research. Admittedly, between Linnaeus and Darwin, before our modern notions of evolution and genetics were established, it was commonplace to subdivide species into subspecies or "races." To some extent, this practice continues to this day. Among animal groups like the birds, taxonomists are fond of creating a new subspecies over a differently colored cheek-patch. Other charismatic groups, like tigers, are similarly divided into distinct subspecies. However, many evolutionary biologists regard these practices with some suspicion, for the following reasons.

At the level of the species, we have a simple criterion for demarcation: the pattern of normal genetic exchange. In particular, the presence of barriers to genetic exchange, be they before or after fertilization, allows us to start classifying organisms into different species, as discussed in chapter 4. But race has no such scientific basis in biology. Before the advent of Darwinism, it didn't matter that taxonomists were somewhat self-indulgent in delineating racial groups. Even Darwin wasn't all that fussy about the distinction between species and races within species. But this situation is no longer regarded as satisfactory to practicing evolutionary biologists. Reading Futuyma's *Evolutionary Biology,* perhaps the standard textbook in the field, you will come across a denunciation of the concept of race or subspecies.[5] The concept has proven considerably more of a nuisance than a benefit.

This rejection of race by evolutionary biology is not based on any earnest ideology. It turns out that it is very difficult to define races for most organisms. The problem is that, within species, local populations typically contain a great deal of variation, rather than conforming to a well-defined type, each in a particular area. Furthermore, there is so much polymorphism in most sexual species that local populations vary between each other in an exceedingly complex manner. One group of populations might have a distinctive coloration, but overlap with other populations in the frequencies of some enzymes. Because of this complexity, there is almost never a single consistent way of classifying the populations of a species into distinct races or subspecies. The end-result has been that evolutionary biologists have not found the race concept worth using. Unfortunately, biologists have not simply expurgated the concept from the field, so some taxonomic races hang on out of historical, academic inertia. And even worse, eugenicists and other ideologues got hold of the race idea and twisted it to their own ends.

An intuitive objection from a nonscientist might be that they can easily see when a person comes originally from Nigeria or Norway.

One is dark-skinned and one is light-skinned, among other differences. But note that this is not the comparison on which the racial issue hinges. "Caucasians" from Sri Lanka can be very dark-skinned, and indeed resemble an Ethiopian to a bus driver in Iowa. Or a Malay from Borneo may resemble a Latino from Costa Rica, to that same bus driver. Or the same bus driver may infer that Tutsis and bushmen must come from a different race, because of their great disparities of height and build. The point is that human populations are differentiated from each other in many complicated ways, ways that are not consistent, ways that are often more superficial than biologically important. The appraisal of the uninformed human eye is often an unreliable guide to the degree or nature of biological differentiation. In particular, skin color, the key test of many casual racial schemes, may be entirely misleading. Population differentiation is a reasonable concept, if overworked by the human imagination, but racial classification of humans is not. Indeed, racial classification is a largely counterproductive concept in all of biology.

The thesis of this book is that Darwinism is an important element of the modern world. But it is not a thesis of this book that its influence has always, or even on balance, been benign. Nowhere is this more apparent than the history of racism. Darwinism tends to have a pernicious effect on debates about biology and polity for two reasons. The first is that it emphasizes patterns of common descent as defining biological lineages. Fish are fish because their ancestors were fish, frogs are frogs . . . , and so on. Pathetically, this is easily extended rhetorically into the erroneous logic that Jews are Jews because their ancestors were Jews, and their ancestors killed Christ, and therefore Jews were bad, are bad, and always will be bad. Likewise, in the nineteenth century, it was commonplace for the English and American establishments to consider the Irish an intractable group given to some type of hereditary predisposition to drink, crime, and unstable family life, a predisposition that was the racial foundation and the inevitable future of the Irish. Thus, the idea of evolution by gradual modification led many cultural and political leaders to characterize despised groups in terms of their supposed racial descent.

The second problem with Darwinism was even worse. The view that there were human races each with their own ancestry and a shared destiny led many biologists and virtually everyone else to the view that human evolution was bound up with the competition of the races. It was not just that the notional Jewish or Irish races were progressing through history, each with its own particular fate. Added to this idea was one of competition, in which the superior races would vanquish—possibly eliminate—the other races. Moreover, this view of history was embellished with the notion that this situation was good, indeed providential, and that the triumph of the superior race(s) was to be seen as a fitting outcome for history.

Most evolutionary biologists don't even want to think about the degree to which Darwinism contributed to the development of racist ideologies in the modern world. They don't really deal with the historical fact that Darwin and Galton accepted the concepts of superior and inferior races, and that Galton was particularly concerned to document the inferiority of the "Negro" and the Australian aborigine. Ernest Haeckel, one of the leading German Victorian evolutionary biologists, paved the way for the development of the elaborate system of German racism that was to develop in modern times. Haeckel was in fact a virulent Aryan supremacist and anti-Semite. Evolutionary biology and racist ideology went, for a time, hand-in-hand.

The irony of racism, and particularly its evil efflorescence in modern times, is that we are seeing the beginning of the end of any possible racist system. The so-called human races are patently absurd. At the molecular level, there is no legitimate way to divide the species up. We are all recently descended from common ancestors in Africa. The biological differences between present-day human populations are fairly trivial, as well as mixed-up like a quilt.

But there is an even more profound problem facing racism. Our fundamental unity as a species is expressing itself in one particularly apt fashion. We are mating with each other. We are breaking down the barriers that existed for tens of thousands of years between populations in Australia, the New World, and the Old World. Population geneticists estimate that it takes only about one successful migrant between populations for these populations eventually to combine into one large genetic unit. We are migrating at a much higher rate than that. The Americas are at the forefront of the reunification of the human species. In the Western hemisphere, representatives of all sizable human populations are present. While the government of the United States and some ideological cranks

maintain the idea of biological race, it is being annihilated genetically. Few American "blacks" have ancestors that are exclusively
African. Intermarriage between individuals of Asian origin and European origin is common, as is intermarriage between "Latinos"
and those of European origin. The Latino group is another case in
point, extensive intermarriage between aboriginal and European
individuals having started centuries ago in South and Central
America. What is your race if your father is a light-skinned "black,"
while your mother is Irish-Mexican? Your race is the human species, and that is the essential situation for all of us. The rest of it is
mostly skin-deep.

Nazism as the Apogee of Eugenics and Racism

Whatever legislative triumphs eugenics enjoyed in the
United States, it was the Germans who took to eugenics with the
greatest enthusiasm. Eugenicists like Wilhelm Schallmeyer and Alfred Ploetz were quick to turn German eugenics into a fanatical
movement, particularly because they, like many others, con-
founded it with the ideas of "racial purity" and "racial hygiene."
German evolutionary biologists had considerable influence on public opinion. With regard to race, Haeckel's theories have been proposed as the templates for Adolf Hitler's more famous outbursts.
The 1937 edition of the Hitler Youth handbook was full of Darwinian theory and genetics, and such science was taken as warrant for
the extermination of Jews. This is not to deny the long-standing
racist elements in German culture. Darwinism did not bring them
into being. But it was fuel for that particular demonic fire. Nor
would it be true to say that all Nazis were reflective evolutionary
biologists. Many of them were just thugs. But as with American
eugenic legislation, which often served the parochial interests of
politicians and their venal allies, Nazi ideologists could turn to the
eugenics movement for scientific credibility.

And they did. Eugenics was an integral pillar of Nazi doctrine.
Adolf Hitler advocated infanticide for all those newborns with
physical defects. Interestingly, physicians were the largest professional group in the Nazi party. The Nazis established a system of

genetic health courts, and physicians had to report genetic disorders to these courts.[6] A "medical eugenics" system in place, the Nazis proceeded to sterilize and euthanize those they regarded as unfit. Their victims included the deformed, the schizophrenic, the mentally retarded, the epileptic, and the institutionalized mentally ill generally. Children who were considered defective were killed by deprivation of care, morphine overdose, or cyanide poisoning, usually without the knowledge of their parents, who would be told that their child had died during medical treatment.

But that was only the "retail" slaughter of the Nazi program, the eugenic winnowing of supposedly defective individuals from the great mass of the European population. The wholesale slaughter was elsewhere, on the racist side of the program. The Gypsies were sent to the camps for annihilation. And then there were the Jews, some of the most brilliant stars in German culture, and vital to the German economy. Because of anti-Semitic theories and myths dating long before the Nazis, the Jews were perfect targets for a "purifying" racism. If eugenic or racist doctrine advocates sterilization, imprisonment, or execution on grounds of biological inferiority, then the widespread anti-Semitic tendency of much German culture at that time naturally led to the program of extermination implemented by Hitler and his willing accomplices.

This particular disaster was not planned or fomented by the eugenics movement generally, at least not directly. But the eugenics program was based on the notion of distinguishing superior from inferior, where only the superior were appropriate breeders for the next generation. The Nazis supplied the thoroughness that led to the Apocalyptic slaughter of so many, even the helpless, newborn, "good Germans" who had birth defects. The Nazis brought forth into the world the darkest side of eugenics, its worst imaginable—even unimaginable—features. But they did not create eugenics. Eugenics was there for them to exploit.

POSTWAR "EUGENIC" PRACTICE

Thanks to the Nazis, there are few movements born in the nineteenth century which are now in such disrepute as eugenics. Essentially all polite discussions of the use of medicine revolve around the alleviation of suffering and the enhancement of opportunity for the genetically afflicted. The treatment of those with genetic diseases is now the opposite of eugenic prescriptions. In this respect, the treatment of infants with PKU disease is virtually paradigmatic. No respectable physician would propose that PKU infants be killed once they are outside the womb. Instead, they are given special diets that enable them to avoid severe mental retardation, as mentioned in chapter 6. Moreover, it is a triumph of American health care legislation that screening for PKU disease is now mandatory for newborns, so that they can receive the medical attention that they need.

The sole remaining "killing field" for eugenics is therapeutic abortion after genetic counseling, when parents are found to be carriers for identifiable genetic diseases. Whether this is seen as hopeful or monstrous depends strongly on religious and other ethical commitments. In particular, there is nothing about being a Darwinian that requires the condemnation or advocacy of therapeutic abortion. In this respect, as in every other, the population genetics of inherited disease do not themselves define any particular ethics of response. Punnett's basic calculation remains valid. Whatever action we might take, it is very difficult for human intervention to change the population genetics of these problems. Most of the genes for these disorders are at very low frequencies, and often they are carried as single copies in unaffected individuals. We are a long way from having the information or the clinical technology to screen people for the full range of genetic disorders which they might carry. For now, little about inherited pathologies is going to change.

AN EPITAPH FOR EUGENICS

In 1925, Herbert Spencer Jennings published *Prometheus, or Biology and the Advancement of Man*.[7] In this brief volume, Jennings argued that the eugenic project is a lost cause. His view was that the deliberate, differential breeding of people would fail, because people are too heterogeneous, because the human environment was often more important than our genetic makeup, and because sexual reproduction shuffles gene combinations to such an extent that the eugenicists would be unable to produce their Promethean human. But his argument had little apparent impact at the time, when eugenics was in its full glory. Now that eugenics is essentially a dead letter, Jennings's little book is something of an appropriate epitaph.

The Promethean ambitions of many Darwinists to direct the evolution of mankind are now virtually defunct. The twentieth century, the great age of meddling ideologies, is almost over. People have had their fill of messianic or scientistic programs for the future of the species. And eugenics was both of those things. It is perhaps a good thing that the dalliance of evolutionary biology with eugenics occurred at a time when genetic technology was so primitive that there was little chance of it having any lasting impact on the species. It is to be hoped that the demonic spectre of Promethean Darwinism is never visited on us again.

PART THREE

UNDERSTANDING HUMAN NATURE

INTRODUCTION TO PART THREE

THIS PART of the book contrasts two major evolutionary theories for human behavior. The first of these theories is based on the idea that human behavior evolves in terms of very specific selective situations, giving actions that can be well understood in terms of evolutionary theory. On this type of model, we are animals with more elaborate behavior patterns than those of other animals. But these are still essentially animal behavior patterns that are shaped by the same Darwinian imperatives, in both human and animal. This type of theory has been called "evolutionary psychology." This theory is based on an essential continuity in evolutionary biology, from animals to humans.

The second theory is that human behavior has evolved to be open-ended and indeterminate, with no more than very general reflections of its Darwinian basis, and often less than that. On this theory, human nature may conform to Darwinian expectations in particular cases, but for reasons different from those that make the behavior of other species conform to Darwinian expectations. Specifically, this second type of theory predicts the general absence of genetically determined human behavior and its replacement by a process of *de novo* cerebral calculation centered on Darwinian fitness. Here, this second theory is called "immanent Darwinism." Immanent Darwinism is based on theories of human evolution that predict substantial discontinuity between animal behavior and human behavior.

These two theories have very different implications for a variety of economic, social, and political questions. In particular, if human nature is not as predictable as evolutionary psychology assumes, then political and other social institutions may be subject to grave instability. Immanent Darwinism supplies reasons for doubting the validity of any deterministic ideology or social-science theory, once technology has developed to high levels.

In the last chapter, the problem of religion is tackled. This is the most important point of contact between Darwinism and the

thoughts and concerns of most people. There is a great deal that
might be said, and in fact a great deal *has* been said on this topic.
Most of it will be ignored here, if only on grounds of volume. The
points that I wish to make hinge on the distinction between religion
as metaphysics and religion as human experience. These two issues
will be handled separately.

8 ORIGINS

From Baboons to Archbishops

WHAT could be more natural than turning to evolutionary biology for a proper scientific story to explain human origins
and human nature? Unhappily, it turns out that human origins are
more murky on an evolutionary accounting than in myths about
mother beaver-gods and father sky-gods. Part of this murk is due
to the technical difficulty of literally unearthing the evolutionary
history of any single species from geological sediment. Another
part is the heavy load of prejudice and hope that humans bring to
their origins. Both of these problems were illustrated by a famous
hoax, the Piltdown Man found in England in 1908. The forged fossil
embodied then-prevalent notions of an enlarged brain leading the
way in human evolution, as if human evolution was guided by
some kind of upward yearning toward an intelligent and noble
condition. This forgery was successful partly because it pandered to
prejudices, and partly because the techniques required to unmask it
weren't fully developed until the 1950s. In many cases, the study
of fossil humans continues to be plagued by a degree of emotional
involvement and acrimony which embarrasses evolutionary biologists who do not work on humans. Modern man has not negotiated
the transition from theological stories to evolutionary stories at all
well, where our origins are concerned.

SAVANNA STORIES

Many tales of human evolution begin with a detailed description of life on the African savanna two or more million years
ago.[1] These stories are usually based on present-day hunter-gath-
erer societies. The males do the hunting, while the females forage

for nuts or tubers closer to the base camp. Such gender-based food gathering is often supposed to select for monogamy, with males provisioning their offspring, going against the normal mammalian pattern of little paternal investment in their young. Selection for such provisioning is then sometimes supposed to lead to the evolution of bipedality (walking on two legs) because that way more provisions could be carried to common dwelling areas. These scenarios are intriguing, full of imaginative interpretations of the ecological and selective constraints facing hominids. There are those for whom this is the real crux of human evolution.

The problem is that we cannot actually know how life was lived on the African savanna one, two, or three million years ago, beyond a few elementary points that are demonstrable from fossil formations. Much of the reasoning about early hominid behavior must be based on hominid skeletal remains, stone tools, and the distribution of these in association with other materials, such as collections of other animal bones. But behavior unfortunately is not preserved in the fossil record.

THE HOMINID FOSSIL RECORD

Though we do not have very direct evidence for changes in intelligence through the hominid lineage, it is uncontroversial that there is clear evidence for significant changes in the brain during hominid evolution. Both brain size and to some extent brain organization can be assessed from fossil skulls.[2] Average brain size has changed quite dramatically over the course of hominid evolution, from around 400 cc. about three or four million years ago to around 1450 cc. for present-day man. Part of this change is expected from the general increase in body size during hominid evolution. However, even when size is allowed for, modern human brain sizes are close to three times what would be expected for an average primate of comparable size.

If it is clear that there has been substantial evolutionary change in the hominid brain over the last four million years, what has been more difficult is determining the pattern of this change. Two major views have developed concerning the pattern of brain evolution: continuous increase or periods of stasis punctuated by rapid increases (punctuated equilibrium model). The evidence probably isn't good enough to decide between them.

Some simple facts about hominid behavior are known from the fossil record. Our ancestors were walking upright about four million years ago. Their teeth suggest an omnivorous diet, made up of meat and plant matter. Our ancestors almost certainly used simple tools made out of tree branches and the like, even back then. After all, today's chimpanzees utilize them, and they aren't upright or any brighter than our ancestors. Our African ancestors had stone tools about two million years ago. These consisted of little more than hand-sized stones with a few flakes chipped off. After another half-million years, stone tools had more elaborate chipping. They were also widespread, from Africa to Europe and Asia. Those are the basic points that are established beyond any reasonable doubt.

Little of this information allows much certainty about the behavior of early humans. We know from cut-marks on animal bones that stone tools were used for removing meat from bones. We know from fossilized piles of animal bone that hominids did transport animal bones to sites to which stone tools had also been taken. Even though tools were used for removing meat from bones, it is not known when they were first used for hunting. It is sometimes assumed that early hominids were hunters, but stone-tool marks overlaying animal tooth-marks indicate that hominid tool users were scavengers at least part of the time. Little is known about the plants in their diet. Archaeology can provide only a very sketchy picture of how life was lived one million years ago.

A Brain Is a Costly Thing to Have

Homo sapiens exhibits truly distinctive elaborations of tool use. The species is now virtually surrounded by its artifacts. Coupled with this rampant tool use is a degree of behavioral plasticity that is unique, evolutionarily. In some respects, each human occupation is comparable to the entire evolved repertoire of behavior of other vertebrates, to say nothing of the much more limited behavioral capacities of invertebrates. This efflorescence of diverse behavior patterns is not based on fixed genetic responses. We have no carpenter social caste, no biological inheritance of occupation from parent to child. Rather, we have the capacity to learn how to accomplish a wide variety of tasks, and to change the repertoire of tasks within our lifetimes. No other organism on this planet has these capacities.

Therefore, at the center of any evolutionary analysis of human behavior must be a well-developed theory explaining how our species came to evolve its behavioral capacities. To operate otherwise would be like analyzing the shapes of cars without regard to how they came to be engineered that way. Just as the design of cars reflects the fact that they are passenger vehicles, so our behavioral attributes must signally reflect their shaping by human evolution. Indeed, some hypotheses about our evolution have radical implications for how we might view our behavior, which will be discussed in the next chapter.

For the non-Darwinian, nothing is more attractive than supposing that the human brain somehow evolved, but that this evolution is no longer particularly important. A common phrase is that "evolution has freed us from biology." That this supposition cannot possibly be correct is revealed by considering the evolutionary mechanisms that have been involved in the origin of our species.

It is an open question how many of the attributes of living things evolve primarily under the aegis of selection. There seems little reason to doubt, for example, that the majority of DNA in the typical vertebrate is not under tight selection. Most such DNA has no function in the coding of proteins, or the regulation of genes. Under these conditions, evolution becomes a random walk in which different DNA sequences are fixed by chance as a result of the accidental predominance of some sequences over others during inheritance. Indeed, any analysis of evolution that does not at least allow for the possibility of such "neutral" evolution is outmoded.[3] Therefore, it is remotely conceivable, before any further consideration, that the human brain, and the behavioral capacities that it engenders, could have evolved by processes unrelated to selection. This in turn requires that any such behavioral features have no associated effect upon fitness, positive or negative. Under this assumption, then, a biologist or psychologist could indeed maintain the view that our behavioral capacities are "freed from biology."

But this assumption is devastatingly refuted by the fitness effects of increased brain size. Leaving aside any presumed benefits to increased cerebral calculation, the mere increase in brain tissue has had pronounced effects on functional human evolution.[4] The main points are straightforward. The large head of the human newborn is a major cause of childbirth mortality, both for itself and for its mother. And that newborn requires extensive care and protection to survive. The female pelvis was substantially remodeled in the evolution of the human, compared to the pelvis of the male. This change in pelvic structure interferes with efficient locomotion, but it helps to accommodate the large head of the newborn. The energetic costs of human pregnancy and lactation are considerable. Brain tissue is also the most energetically costly tissue from the standpoint of basal metabolism. Up to 40 percent of human basal metabolism may be used to fuel the brain. All of these attributes reflect the action of evolutionary mechanisms very different from mere neutrality. The human brain cost a lot to evolve. It cannot be a neutral attribute.

WHY DID WE GET SMART?

Since the human brain was not an evolutionary "freebie," it must have been produced by selection. This point is full of significance. If our brains have evolved because of some strong selective force, then the nature of that selective force, or forces, becomes central to the riddle of our present natures. From different hypotheses about the selection that led to our evolution must come radically different consequences for the theories that can tenably explain our behavior. We now turn to these alternative hypotheses.

As we discussed earlier, savanna stories for human evolution combine elements that are irrefutable and elements that depend on specific historical events. In other words, these scenarios combine the deficiencies of being both overly speculative and overly fragile. As such, we must reject that approach to understanding hominid evolution, for our present needs. Our goal should instead be to consider those ideas that are not deficient in those particular ways, whatever their other deficiencies.

There is first the question of the target of selection. For many adaptations, this is difficult to discern. In the case of the evolution of man, fortunately, there appears to be wide agreement on two important targets for selection: "technical intelligence" and "social intelligence." Both of these may or may not involve selection for enhanced tool-use, better verbal communication, and so on. Indeed, the range of specific possibilities for these two modes of selection is substantial. What is of interest here is how to differentiate them coherently.

By technical intelligence, I mean capacities that pertain to food acquisition, predator defense, and so on. The situation is that of Robinson Crusoe before his man Friday, the solitary tool-user, perhaps with immediate family, coping with the struggles of existence, where *the adaptation in question enables the individual to better exploit or survive the environment* through the use of learned behaviors.

By social intelligence, I mean capacities that pertain strictly to interactions with members of the same species, where such interactions may be violent, cooperative, or something in between. This is the domain of the game-player, manipulator, or deceiver. In this case, *the adaptation in question enables the individual to better exploit or survive the actions of members of the same species* through the use of learned behaviors.

As artificial as this dichotomy may seem, it separates the vast majority of hypotheses that have been put forward to explain human evolution. We now turn to an outline of some of these alternatives.

SELECTION FOR TECHNICAL INTELLIGENCE: TRIUMPH OF THE NERDS

Many animal species use tools, from the termite-fishing twigs used by chimpanzees to the egg-cracking rocks used by Egyptian vultures. In the vast majority of species, there are only a few tool-using techniques, and in many of these cases the tool-using is genetically encoded, as in the nest construction of social insects. Indeed, leaving aside the edifices and technologies used by social insects, almost all animal tool-use is at a very low level of sophistication. And it is significant that the most dramatic animal engineers, the social insects, do so on a genetically programmed basis. The only known evolutionary lineage in which tool use is enormously elaborate and also overwhelmingly learned, as opposed to encoded genetically, is the hominid lineage. Other primates, most notably chimpanzees, have some degree of learned tool-use. But such tool use is ancillary to most aspects of the ecology of these other great ape species.

Given the very obvious differences in material culture between humans and their closest primate relatives, it is perhaps not surprising that the idea of "man, the tool-maker'" has a long history. Even before Darwin's theory of evolution by natural selection had been accepted, a number of authors, such as Benjamin Franklin and Thomas Carlyle, had drawn attention to man's special affinity for tools. Of course, at that early point in thought about the subject, the fact of tool use only served to set humans apart from the animals and was not raised as a mechanism for human evolution, for no such mechanism was sought.

One of the first to speculate on the role of tools in human evolution was Friedrich Engels, who in 1876 wrote an essay entitled "The part played by labour in the transition from ape to man."[5] Engels argued that the freeing of the hands by the adoption of a bipedal posture led to the mastery of nature through manual labor. The development of labor, in turn, "necessarily helped to bring the members of society closer together. . . . [M]en in the making arrived at the point where they had something to say to each other" and necessity created speech.

As the picture of human ancestry started to fill out during the middle part of the twentieth century, further speculation on the importance of tools in the evolution of intelligence appeared, although it was some time before theorizing went much beyond what had been said close to a century before. Kenneth Oakley's 1959 description of prehistoric tools noted that tools made man "the most adaptable of all creatures."[6] Like Engels, Oakley thought the adoption of bipedality would have freed the hands for the manufacture and manipulation of tools. Although he believed that the real difference between ape and man lay in mental capacity, he suggested that tools may have been responsible for increasing human mental powers.

Oakley did not offer anything in the way of a theory for how tool use may have led to an increase in human mental powers. Sherry Washburn, riding the crest of a wave of significant new discoveries in paleontology, was willing to go somewhat further in his 1960 speculations on the role of tools in hominid evolution.[7] He argued that positive feedback for tool use led to consistent bipedality, resulting in a whole new way of life and further selection on many parts of the body, including teeth, hands, pelvis, and brain. The increase in brain size related to tool use necessitated further adaptations as human infants had to be delivered at an earlier stage of development, thereby requiring increased maternal responsibility, and decreasing maternal mobility.

Washburn's ideas illustrate the now-commonplace assumption that the significance of tools was in greatly enhancing the capacity for procuring and processing food. As Washburn says, "The huge advantage that a stone tool gives to its user must be tried to be appreciated. Held in the hand, it can be used for pounding, digging or scraping. Flesh and bone can be cut with a flaked chip, and what would be a mild blow with a fist becomes lethal with a rock in the hand." Tools can also be used for fashioning other more specialized tools, and for creating such artifacts as containers, essential for carrying and storing excess food.

SELECTION FOR SOCIAL "INTELLIGENCE": THE MENTAL ARMS RACE

Evolutionary change does not come about only from individuals having attributes that allow better exploitation of a particular ecological niche; it also comes from individuals competing better with other members of their species, a main theme of chapter 3. That is, evolution is not only adaptation to the material environment. It also involves selection on attributes that determine success in interactions with members of the same species. Other hypotheses about hominid evolution have focused more on this selective pressure. Our interest in reviewing these hypotheses is how interaction between individuals might produce the selective pressure for the changes in the hominid brain observed over the last five million years. Note that some of these hypotheses assume that technical intelligence had already evolved, to some extent, before selection on "social intelligence."[8]

Competition is generally assumed to provide the context for fitness benefits from improving social intelligence. There are a number of domains in which such competition might operate. First, there are intersexual relations, whereby male and female negotiate the provision of mating opportunities. Second, there is intrasexual competition among males regarding access to females. Competition for resources, such as food and territory, provides a third context in which social intelligence would be of value. Clearly all three of these domains are ones which are of general relevance to most animal species.

It has been supposed, however, that certain conditions in hominid evolution, such as surplus food, cooperation, or tool use, changed the conditions of competition. A surplus of food available from hunting or scavenging may have allowed males to provision females with food in exchange for mating opportunities. Such exchanges would be ripe for exploitation by cheaters of both sexes. Similarly, if cooperation became an important part of hunting and foraging, among other areas of hominid activity, then it is clear that cheating would again be a possible strategy. For example, a successful hunter might not share with another who had helped in the hunt. The advent of tools, especially for hunting, would have had serious implications for combat within the species. Such combat is believed to be a good candidate for the evolutionary origins of the large human brain, because of the intense selection pressure that it would produce. In conflicts of this kind, there would be clear advantages for creative thinking and planning for the individual warrior, or indeed the soldier in proto-military combat. An attractive feature of this theory is that it is clear that varying degrees of success in armed combat could lead to large variances in fitness, and thus effective selection, thanks to death, castration, and other misadventures of battle. While such violent competition may provide selection for social intelligence, this does not mean that all social intelligence would be concerned with violent competition. Social games may be more easily won if an ally can be recruited to your side. It is notable that primate social behavior between two individuals is affected by the presence of a third. Without question, multiplayer social games will be far more complex than two-player games, in that the amount of information that has to be processed is considerably greater and the set of options correspondingly enlarged. In addition, individuals will have to be able to keep track not only of other individuals' relationships to them, but also the interrelationships between the other members of the group.[9]

But this is only our starting point. Consider the problem of two stags fighting for a mate. This kind of evolutionary situation is now usually analyzed in terms of the alternative strategies discussed in the chapter on selection: Hawk, Dove, Bourgeois, etc. These are all candidates for unbeatable strategies. Unbeatable strategies arise from the particular evolutionary situation faced by members of each population. Thus, the determinants of the unbeatable strategies for the stags will include how many deer there are to mate with, how sharp the antlers of the other stags are, how fragile antlers are, and so on. Most of these factors will be determined by general aspects of the morphology, ecology, and physiology of the organism, which cannot be changed by the contesting stags. In other words, organisms can't usually "cheat" in their evolutionary games, because the game rules are set by the basic biology of their species. This is why evolutionary game theory can work. Evolutionary games are normally static, and behavior evolves within the context of stable rules, much as chess-playing has evolved for the last few centuries, during which the rules of the game have not changed.

That is the situation for most animals, but not for hominids that can use tools proficiently. If a species evolves within a "learned tool-use" niche, then the deadliness of its violent conflicts will not be determined only by the basic biological constraints faced by the species. In particular, if a flexibly tool-using species develops hunting weapons that are sharp and lethal, then the employment of those tools will determine the outcome of evolutionary game situations. But, unlike antlers, horns, claws, or fangs, these weapons would not be built-in, would not be genetically encoded. As such, they would not define a stable evolutionary game.

Now this is not to argue that evolution would then come to depend on building better and better tools, though that may be some part of the story. The point here is that *potentially deadly weapons that are not genetically based undermine the consistency with which any particular evolutionary game strategy will be an unbeatable strategy*. As such weapons become more important, partly because their use becomes more versatile, then selection for normal unbeatable strategies will be undermined. The validity of evolutionary game theory will break down, for lack of sufficient stability in the evolutionary games. The old patterns in the evolution of social behavior will no longer be sustained. The stage is set for a completely new kind of selection.

If you are always confronted with new games, and their solution determines your Darwinian fitness, then what game strategy is going to be successful? The most successful game strategy will be one that is determined by direct calculation, including experimentation, in each game situation. That is, the most successful strategy cannot be genetically specified. This may be contrasted with the evolution of social behavior in most animals, which apparently follow simple genetic rules of retaliation, territory, and the like. Those organisms are probably following strategies that are close to unbeatable strategies, and adherence to a fixed strategy is the best approach for Darwinian fitness. That these strategies do not depend on the overall level of neural complexity is revealed by the remarkable parallels in contests over mates in a wide range of organisms, from butterflies to monkeys, described in chapter 3.

The key point is that learned tool-use will eventually select for the direct calculation of social strategies. On some universal, social intelligence, selection theory, it might be thought that all kinds of species should evolve social intelligence. It might always seem better to base one's behavior on more information and more inferential sophistication. Evolutionary game theory, on the other hand, suggests that the best strategy will often be rather simple, and the empirical finding of retaliation and other simple strategies throughout the animals suggests the validity of evolutionary game theory. It also suggests the dubiousness of the general invocation of selection for social intelligence.

But the hominid lineage would be an entirely different case. In our lineage, good "dumb" unbeatable strategies could not be the best strategies, for the simple reason that no fixed strategy was likely to suffice. The selectable character becomes "strategy-calculation." And not only will there be selection for strategy-calculation, there will be selection against the adoption of fixed genetic strategies. The reason for this is that, if fixed genetic strategies aren't the best strategies and other players have strategy-calculation, then they will often be able to exploit or counter any players using genetic strategies. Thus, there should be selection for social versatility and the erosion of stereotyped social behavior.

To return to the evolutionary process, selection on hominid strategy-calculation can be fairly characterized as a mental arms race.[10] The key factor is the relative capacity for strategy-calculation of each player. Those that can calculate further, in more detail, considering more alternatives, would have the advantage. In terms of each individual contest, this is like a human game, such as chess, and all the same general features of tactical investment would still apply. But the overall problem of strategy-calculation is *also* frequency-dependent, it also has "game" features. In game-theoretic terms, selection on strategy-calculation is a "bidding" situation, which can be thought of in the same general way as a fine art auction. The highest bid wins, the lower bids lose. A major *disanalogy* with human auctions is that, in the case of evolutionary games, the bids involve the deployment of real resources. These resources might be the energetic cost of protracted calling in male birds, the predation risk incurred by calling male toads, or the costs of growth of antlers in stags. Not all these costs are recouped by those with lower "bids." Similarly, for selection on strategy-calculation, there must be investments involving growth of brain, programming of brain functions, and the calculation time itself. All these will be expensive, in terms of resources that, in principle, might otherwise have been allocated to other Darwinian functions, such as growing muscle tissue or digestive tract.

This situation has been examined mathematically and the results are surprising. In mental arms races, selection tends to smear the distribution of investments in intelligence. It does not tend to produce the pattern of evolution that humans have exhibited, with large increases in brain sizes among almost all members of the species compared to our ancestors three million years ago. Even if we make a generous allowance for possible complications of genetics, mathematical analysis predicts that typical human brain sizes should be similar to those of early hominids, and the entire range of brain sizes, from that of three million years ago (400 cc) to that of the present (1450 cc), should occur in contemporary human populations. As our present-day, brain-size distributions are nothing like this, qualitatively, pure mental arms race selection arising from tool use is not a plausible model for human evolution. Normal brain size has increased too much. Moreover, this result does not depend on the verbal particulars of one or another "social intelligence" model. All these models require brain investment, and they are all arms race models. None of them are viable, so long as they are strict "social-intelligence" models.

THE MENTAL ARMS RACE AMPLIFIER: THE WHOLE HUMAN?

In the foregoing analysis of the mental arms race, it is supposed that there is a cost to the acquisition of extensive calculating capacities, a cost that cannot be mitigated. This assumption is almost certainly erroneous, in that an extensive strategic calculating capacity would also be useful in other biological contexts. One of the most obvious of these other contexts, since it is functionally analogous to armed combat situations, is hunting. Outwitting the prey would be infinitely easier for a hominid product of mental arms race selection. And thus there must be some sense in which the "costs" of the mental arms race can be "paid" in ecological advantages.

One way of thinking of this evolutionary theory is that it describes a transition to a full "learned tool-use" niche, where the initial "colonizers" of the niche are inefficient. Later exploiters of the niche would evolve much more efficient exploitation of the niche, and thus of tools, and so derive greater fitness-benefits. This is the "positive-feedback" idea in the evolution of tool-using. But eventually, at some level, the costs of an enlarged brain, protracted child learning, and so on must exceed the further gains obtained from more efficient learned tool-use. Beyond that point, selection for technical intelligence will not succeed in increasing the character much further.

But if one supposes that a mental arms race begins before this intelligence level, and selection acts on a general computational ability, *usable in both technical and social contexts,* then such "amplified" selection would rapidly increase this computational ability, for some time. Inevitably, once this ability is costly outside of social competition, a normal arms race will occur, at cost, smearing the distribution of calculation abilities. In this scenario, evolution has three phases: (*i*) selection for technical intelligence, only; (*ii*) selection for both technical intelligence and social intelligence; and (*iii*) selection for social intelligence only. In phase (*ii*), selection for technical intelligence acts as a *mental arms race amplifier,* because it frees the arms race from costs. It is during this phase that evolution of increased brain sizes and other adaptations should be particularly rapid.

Suppose that a mental arms race starts once selection for technical intelligence alone has created sufficiently efficient learned tool-use. This would be the point at which lethal tool-use has destabilized hominid evolutionary games. From that point on, we assume that mental arms race selection continues. Initially, it will be amplified by concurrent selection for tool use in an ecological context. But after some point, the mental arms race will have to be paid for. Further increases in investment in strategy-calculation will be costly. However, the mental arms race will generate a smear in levels of general intelligence, perhaps reaching a threshold above which the ecological benefits again exceed the costs. Once that point is reached, the arms race is again paid for by the benefits in the ecological arena. This process can continue, with periods of strict arms races bridging gaps between the "peaks" of ecological benefit, on which the mental arms race is amplified. Increases in general-purpose calculation will come to an end only when the mental arms race never again attains a peak of "technical intelligence" benefits. In other words, the process runs out of steam when no further mental arms race amplification is available.

WHY WE AREN'T LIKE THE OTHER ANIMALS

There are a number of features of this theory that need to be made explicit. The first is that it only works for selection on brain functions that are sufficiently general that they can undergo positive selection in both technical and social contexts. This is an important feature of this model, as will be discussed in the next chapter. But for now it should be emphasized that only sufficiently general calculating mechanisms could be favored by the mental arms race amplifier mechanism. In effect, these mechanisms would have to be generalized intelligence mechanisms, rather than specific mechanisms for social or technical functions. The kinds of brain adaptations that this theory predicts are those that produce a calculation capacity that can serve almost any fitness-relevant goal.

The second point is that this combination of selection mechanisms should be able to give rise to very rapid evolution of increased brain size, as the brain takes on more and more general functions. This situation arises because the arms race for brain functions is not only "paid for," but actually amplified by selection in the ecological arena. Most other organisms pay large fitness costs for their behaviors and structures that pertain to competition, from conspicuous male coloration and mating behavior to the structures and behaviors used in male combat. The hominid lineage that gave rise to *Homo* may have experienced a situation in which the evolution of an adaptation for rivalry was often free of net costs.

This combination of selection mechanisms can also give rise to much more protracted evolution of general brain functions than either could singly. This protracted evolution may be a large part of the explanation for the great elaboration of the human brain. This is important because the expansion of the human brain over the last two million years is one of the most rapid and sustained such morphological developments known in the fossil record. This evolutionary record is one that requires an unusually powerful selection mechanism. The mental arms race amplifier has that quality.

A third point is that the specific curve for the fitness-benefits of technical intelligence will depend on the environment. But the superimposition of a mental arms race will blur any such environmental dependence in the evolutionary outcome for several reasons. In the absence of the mental arms race, the evolutionary equilibria of populations selected *only* for technical intelligence in different habitats should be quite different, depending on the specific brain sizes for which the fitness costs become significant. But the mental arms race pushes the evolutionary trajectory from peak to peak. The mental arms race equilibrium distribution is also a "blur" or "smear." The net effect of the mental arms amplifier mechanism will be to obscure any relationship between specific environments and levels of generalized calculation ability. Thus, for example, this theory has as a corollary the absence of well-defined, stereotypical, differences in fitness-calculation between human populations that have been long-established on different continents, the so-called "races."

ALTERNATIVES FOR GENESIS

In the end, we can differentiate a range of alternative stories for the evolution of humans. These stories are divided into two broad classes. The first class consists of those stories based on some specific combination of selection pressures, pressures that are thought to have produced a number of specific human abilities and propensities in the course of our evolution. Most of the human evolutionary stories in the popular literature and on television are of this kind. This kind of story implies a particular type of human nature, one that can be viewed as an elaboration of universal mammalian patterns.

The second class of evolutionary stories about human origins is based on general selection mechanisms, such as selection for technical or social intelligence. This type of story predicts a different kind of human being, one in which animal patterns of behavior are evolutionarily corroded, in whole or in part. The contrast between these two broad visions is a major concern of the next chapter.

9 PSYCHE

Darwinism Meets Film Noir

THE ISSUE at stake is the relevance of Darwinism to the understanding of human behavior. Given the premise of Darwinian evolution, what can we infer about the nature of the human mind and the behavior that it produces? Given the difficulty of reasoning from basic neurobiology to our complex behavior, Darwinism might be an alternative foundation for understanding human nature. Or it might not.

Mainstream evolutionary geneticists have been running away from this issue for decades, despite the great interest of nonspecialists. Recently, it has been zoologists and anthropologists who have wanted to explain human behavior using Darwinism, despite the misgivings of most evolutionary geneticists. Inevitably, this set the stage for a mighty clashing of doctrinal tribes.

SOCIOBIOLOGY

The first sounding of trumpets came in 1975. E. O. Wilson, a Harvard University entomologist and population biologist, published a book entitled *Sociobiology: The New Synthesis*. In this book, he proposed that the study of behavior, human and animal, be recast in terms of modern evolutionary biology. In support of this proposal, he provided a synoptic overview of the behavior of a vast range of animal species, interpreted in a Darwinian manner. In spite of the claim of novelty, Wilson's program of research was nothing new. Darwin had the same ideas around 1840, and he published many of them in his *Descent of Man* and *Expression of the Emo-*

tions in Man and Animals. Since Darwin, behavior has been a frequent topic in evolutionary biology, and indeed evolution has frequently been discussed by those who study behavior.

In *Sociobiology* and again in *On Human Nature*, Wilson argued for the interpretation of human behavior in terms of his unified science of behavior, sociobiology. In particular, Wilson argued for the application of evolutionary theories like kin selection to human behavior, in order to explain such behavior as nepotism. The true novelty of sociobiology was that Wilson had better academic credentials than earlier proponents of the Darwinian explanation of human behavior. Even a German Nobel Laureate, the behavioral zoologist Konrad Lorenz, was not such a threat for modern Darwinism, despite his *On Aggression*, which explicitly addressed such subjects as human violence and war-making. Lorenz was offering his individual views, based primarily on his studies of animal behavior. Wilson's sociobiology was an entire scientific program. Human behavior was now to be studied like the behavior of fruit flies and geese, with a persistent effort to resolve or reveal all the Darwinian motives inherent in it. Wilson was interested in total explanation, not picking out a few oddities or peccadilloes for the purposes of intellectual *frisson*, as Desmond Morris had in *The Naked Ape*. Wilson's sociobiology was to mark the end of a recessive Darwinism, a Darwinism polite enough to coexist peacefully with traditional social science in the tranquillity of the modern research university. Wilson's approach was a threat to the *pax academica* between biology and the humane studies.

CRITICS OF SOCIOBIOLOGY

By academic standards, pandemonium broke loose.[1] E. O. Wilson was denounced from the pulpits of modern academia and literary pretension. The *New York Review of Books*, heavily influenced by Richard Lewontin—the leading population geneticist—began an attack of unparalleled vigor.[2] Echoes of this attack may be read on its pages to this very day. Numerous critiques were published in lesser organs of publication, so many that they have been collected together in book compendia. Wilson was attacked by chanting students and ideologues, even doused with water on one occasion. It was insinuated that he was a racist, a liberal, a lackey of capitalism of one kind or another, and so on. No Darwinist had been treated to such sustained moral and political denunciation since Charles Darwin himself.

One of the widespread criticisms of Wilson was that he was guilty of "biological determinism," the doctrine that everything about human behavior is determined biologically, particularly genetically. This sort of doctrine, whether right or wrong, suffers from guilt by association with the race-supremacist theories of Nazis, Ku Klux Klansmen, and the like. It harkens back to the bad old days of eugenics, described in chapter 7, the last time Darwinists had become involved with "social" questions, to their great regret. However, there is little evidence that Wilson advocated any such total biological determinism. Indeed, the main attack against Wilson was that, in suggesting *any* hereditary component to human behavior, Wilson was letting biological determinism slip into the study of human behavior by the back door. This, it was argued, would then pave the way for racist and other forms of biological determinism.

The problem facing this line of criticism is that it renders problematic the evolution of human behavior in the first instance. If there is no genetic component to human behavior, then why does our behavior differ from that of chimpanzees? Some genetic basis for human behavior is inevitable on any reasonable scientific theory. Therefore, hysteria about introducing the problem into discussions of human behavior only tends to undermine the rejection of extreme biological determinism by confusing such determinism with any evolutionary or genetic hypotheses about human behavior. The criticism inadvertently does exactly what it rejects. By stigmatizing any invocation of genetics or biological evolution, and associating such views with racism and the like, it fosters the credibility of such odious ideas. Gene-free, evolution-free behavior is essentially an absurdity. All behavior must arise from a biological substratum, however unspecified the contents of that behavior are genetically. The critics attacking Darwinians on these points have nothing to fall back on, unless they are very strong fundamentalists, and can invoke God. Marxist, and so materialist, critics of Wilson's program of research only strengthened his case, and weakened their own, with their invocation of biological determinism.

THE LARGER PROBLEM OF ADAPTATIONIST
REASONING

A more cogent criticism of sociobiology was provided by Gould and Lewontin in a paper where they compared biologists of Wilson's type to Dr. Pangloss from Voltaire's *Candide*.[3] Dr. Pangloss was the pedant who could explain everything from syphilis to earthquakes in terms of all things being for the best in this best of all possible worlds. And these explanations were terribly lame, in Voltaire's story. The function of the nose, for example, was supposed to be the propping-up of eyeglasses. Voltaire's target was a vein of fatalistic philosophy that was common in the eighteenth century. The target for Gould and Lewontin was adaptationist stories that mirrored Dr. Pangloss's just-so stories in their assumption that all characters had to be for the best in this best of all possible worlds. Their point was that if you could shoot down one of the sociobiology stories by means of careful experimental study, the sociobiologists would just invent a new adaptive explanation. (This problem has already been discussed in chapter 3.) And as bad as such a lack of intellectual inhibition must be where animal behavior is concerned, its potential for mischief in the explanation of human behavior must be still greater.

This was a telling criticism. To some extent, it remains so to this day. However, one of the major developments in evolutionary biology since Gould and Lewontin wrote their paper is a widespread appreciation of the difficulty of understanding adaptation. But the difficulty of a scientific objective is not a warrant for its abandonment. All the criticisms of Gould and Lewontin can be accepted, but despair and intellectual nihilism are not the inevitable sequel. There are comparable, or greater, difficulties with understanding such problems as speciation, the evolution of sex, and the origin of life. Admittedly difficult problems like adaptive behavior in hominid species should be approached with attention to their grave difficulties of theoretical explanation and experimental testing. But if enough care is taken, there is no fundamental scientific reason why human behavior cannot be discussed by evolutionary biologists. Individuals may have a strong religious or ideological commitment that prevents them from pursuing a Darwinian approach to human behavior, and their self-restraint is their concern. But no such strongly felt commitment of some individuals warrants their censorship or physical humiliation of a scientist who is merely, if naively, pursuing a Darwinian approach to the subject.

On the other hand, even if the witch-hunt against sociobiology in the 1970s was unwarranted and overblown, that is no demonstration that Wilson's sociobiology program has more credibility than earlier searches for phlogiston or cosmic ether. Even when a scientific theory or approach is conceivable, it still may not be valid. However, sociobiology does not exhaust the possibilities for the Darwinian analysis of human behavior, as we will see.

SPECIFIC PROBLEMS WITH HUMANS

There are serious and general empirical problems that need to be resolved in the study of human behavior from the perspective of evolutionary biology. In particular, human beings erect systems of societal norms for behavior that sometimes do and sometimes do not conform to expectations that some scientists have derived from evolutionary theory. One example of each will be given here.

One of the very few cases where a direct sociobiological analysis of human behavior has been successful is incest avoidance.[4] It is a basic genetic phenomenon that inbreeding in normally outbreeding sexual species leads to the production of defective offspring, as outlined in chapter 6. This is called "inbreeding depression." Specifically, mating brother to sister or parent to child is expected to lead to high levels of inbreeding depression. Natural selection should act to prevent such behavior. In many mammal species, inbreeding avoidance is common. Incest taboos are common among known human societies, although incest has occasionally been encouraged between high caste individuals, such as the Pharaohs of ancient Egypt. Incest avoidance appears to be genetically determined, rather than taught. Unrelated children reared together in kibbutzim are strongly encouraged to marry each other, but in fact rarely do so. Likewise, if Taiwanese child brides are adopted into the groom's family when both are infants, the marriages are rarely successful. Often the partners in the arranged marriage feel little sexual attraction, if not active revulsion, toward one another. What seems to be involved here is a kind of "imprinting" behavior, in which infants reared together come to regard each other as siblings, and therefore not appropriate mates, regardless of either the actual biological situation or social reinforcement. This story fits closely with a sociobiological perspective, in which specific features of human behavior are supposed to have been shaped by evolution so as to maximize Darwinian fitness. This might seem like a wonderful start for Wilson's program of research and analysis. But it is also the end of this program. No other feature of the human psyche is quite so amenable to explanation in terms of evolutionary genetics.

Some of the typical problems are illustrated by attempts to explain human behavior within families and tribes in terms of kin selection.[5] One of the typical expectations derivable from kin selection theory is that human beings should systematically favor their biological relatives in economic and political transactions. It is propitious for this type of theory that such behavior is recognized by such words as "nepotism" in English, and many similar words in other languages. On the other hand, as shown by the anthropologist Marshall Sahlins, nepotism is not associated with biological relatedness in the strict fashion suggested by kin selection theory. In many human populations, kin attributions do not correspond to true genetic kinship. Often unrelated individuals are adopted as kin, even when biological kin are present. The adoption and succession of later Roman Emperors illustrate this kind of supposedly nonadaptive behavior pattern. The Romans did not tend to follow the later European practice of inheriting office biologically. Such behavior makes little sense in terms of simple sociobiological reasoning. On close inspection, nepotism and kin selection are not the perfect match that a sociobiologist might suppose.

In the case of incest avoidance, sociobiological reasoning seems successful. Notably, this is a case in which the fitness consequences of biologically specific behavior are overwhelmingly predictable. When we turn to nepotism, sociobiological expectations already begin to break down, and major deviations from them are obvious. Beyond nepotism there is a vast range of behavior, practically all human social behavior, that does not fit sociobiological models in any clear way. These findings suggest that while evolutionary biology may have something to contribute to our understanding of human behavior, it is unlikely that it is something as simple as the direct extension to man of theories found to be successful with ants, bees, and termites. Sociobiology was an interesting idea that doesn't have much in the way of legs on which to carry us.

Even E. O. Wilson himself realized that much had been left out of his original "synthesis." His solution to this problem was to bring in "culture," the transmission of learned information from person to person. In particular, with Charles Lumsden, he developed an elaborate theory for the coevolution of culture and genetic learning predispositions.[6] This theory was controversial within scientific circles, even if the general public paid less attention to it. One of its distinguishing features was that it seemed to focus on such utterly trivial features of culture as the height of hemlines, while neglecting the Darwinian focus on reproduction and survival that had made sociobiology so plausible in the first place. This general theme of the joint processes of cultural evolution and genetic evolution has received attention from other scientists as well, at least at the theoretical level. It is unclear how much this highly mathematical work has contributed to our understanding of specific human behaviors. Compared to the original, rather simple, ideas of sociobiology, this body of gene-culture coevolution theory is extremely difficult to test, at least in humans.

EVOLUTIONARY PSYCHOLOGY

"Sociobiology is dead, long live evolutionary psychology!" If sociobiology was E. O. Wilson's naive attempt to unify all biological and social sciences, then evolutionary psychology is its heir. Particularly nourished by anthropologists like Jerome Barkow and the team of John Tooby and Leda Cosmides, evolutionary psychology has been much more sophisticated than sociobiology.[7] For one thing, these people know something about the social sciences. They are social scientists. They even know things about people, both in industrialized societies and in those that are considered primitive. They have many hours of anthropological fieldwork between them.

Thus, evolutionary psychology takes the simple Darwinian ideas of sociobiology and harnesses them to relatively recent concepts from anthropology. They have also been considerably less provocative about their relationship to the social sciences, and considerably more sophisticated about ideological issues. As such, evolutionary psychology has been a more elusive beast, less of a provocation that would motivate the academic blood-lust that brought down sociobiology. Again, this does not make evolutionary psychology any more or less likely to be true or useful, from a strictly scientific standpoint. It primarily makes evolutionary psychology more formidable as an academic competitor.

In some respects, evolutionary psychology has successes to claim. If one evaluates evolutionary psychology compared to theories like Freud's psychology, which is a more common maneuver now than in sociobiology's time, then evolutionary psychology is far more satisfying. An early move by evolutionary psychology was to cannibalize the successes of sociobiology, and then use them in its contests with other intellectual movements. It is credible that humans avoid incest because of selection against tendencies to have coitus with close relatives, given the deleterious genetic consequences of extensive homozygosity. The alternative is mumbo-jumbo about Oedipal complexes, repression, and denial, a farrago of Viennese nonsense. Likewise, the widespread sexual preference of men for healthy young women of adequate pelvic and pectoral anatomy for child-bearing is plausibly related to basic Darwinian selection for reproduction, instead of some arbitrary aesthetic accident. And such sexual selection has been a major weapon for evolutionary psychology.[8] Another theme has been patterns of murder, particularly its dependence on genetic relatedness.[9] Evolutionary psychologists are not shy about pressing home their major advantage, that of being Darwinians, against social scientists who have scorned biological evolution for decades, sometimes even to their own regret.

Immanent Darwinism

The strongest criticisms of evolutionary psychology must come from a Darwinian standpoint. The greatest source of credibility underscoring evolutionary psychology is the attempt of its practitioners to operate from Darwinian foundations. It is counterproductive to criticize these latter-day sociobiologists by attacking their invocation of Darwinism. What other general theory is there that provides a demonstrably better foundation for understanding the organization and functions of human behavior? If there is to be an alternative to evolutionary psychology, what is required is a Darwinian analysis that shows why and how the favorite devices of evolutionary psychology won't work.

One such critique follows naturally from the mental arms race amplifier theory sketched in chapter 8. One of the key features of evolutionary psychology is the analysis of the evolution of human behavior in terms of the genetic evolution of highly specific behaviors. To a first approximation, this is an excellent approach to many features of animal behavior, which often exhibits specific genetically determined characteristics that probably evolved by natural selection.[10] Optimal foraging theory is the type-specimen of this kind of reasoning. In optimal foraging theory, the relevant food objects are characterized in terms of their nutritional value and their distribution through space and time. The searching or waiting behavior of the forager is also quantitatively characterized, and the key allocative possibilities established. Thus, a nectivorous bird has an optimal foraging situation defined, in part, by the distribution of flowers in its habitat, the caloric value of their nectar, the costs of flight from flower to flower, and so on. From such information, calculations can be made concerning the ideal flight patterns for the enhancement of the bird's fitness. The Darwinian prediction is then that these birds will conform to these flight patterns, thanks to genetic evolution arising from natural selection. Such behavioral analyses are at least somewhat successful, though there are still arguments about cases where the initial predictions are not borne out. Nonetheless, this type of scientific study has at least some cogency to it. This is the type of behavioral analysis that evolutionary psychologists seek to emulate when they study humans.

But some serious thought about the human case immediately raises problems. Do humans in supermarkets follow optimal foraging patterns, like nectivorous birds in tropical rainforests? If so, is that because we have evolved appropriate foraging strategies in the fifty years since supermarkets became common? That's just two generations. Any basic knowledge of population or quantitative genetics disabuses one of that notion. There hasn't been enough time. Yet, when placed in these evolutionarily unprecedented situations, humans do many novel things in a quasi-functional, pseudo-Darwinian fashion. That is, we frequently behave *as if* we have evolved to produce a particular type of behavior that is optimal for Darwinian fitness. And we do so even when that is impossible.

One interpretation of this situation is that humans have hardly any genetically encoded Darwinian tactics appropriate to particular behavioral contexts. Instead, we could have evolved functional, but indeterminate, behavior because of the mental arms race amplifier. On this model, behavior is determined by an immanent process of calculation taking place in the brain, not by genetic evolution arising from natural selection. There is a great deal of information built into the determination of behavior in, say, social insects, the focus of E. O. Wilson's research. But much of *that* information is supplied by natural selection acting on genetic variants. In the vast majority of animal species, which are simple invertebrates, behavior is almost entirely genetically programmed. The programming of behavior is *not* immanent for most animal species. But for a species with open-ended neural calculation, the programming of behavior will be largely immanent, not genetic.

A second major feature of behavior in organisms of this type is that it will not be orchestrated in terms of a large set of highly specific behavioral routines. That is, there will be a unified aspect to the mental process. In a sense, this could be described as saying that members of such species have "a mind." The calculations that shape behavior undergo some centralized clearing, rather than individual mechanisms being triggered by specific stimuli in a genetically canalized manner. Thus, internal processes of calculation would be elaborate and self-reflective. Such patterns as "self-awareness" would be typical.

A third predicted feature of human behavior, according to this model, is that it should focus on fitness ends. While there would be a generalized mental process shaping behavior, rather than specific genetic programs, this generalized process would nonetheless be oriented toward anticipated fitness-contingencies. That is, while this hypothetical organism would be free of most genetic specification of behavior, it would not be free of an overarching Darwinian imperative. The behavior would exhibit immanent Darwinism.

Fourth, the behavior of a product of mental arms race selection would not be easily interpretable, in most instances. Since "social intelligence" selection will strongly favor deceit and dissimulation, even though behavior may be attuned to some fitness goal, it should not be as obviously so attuned as, say, two male horned beetles fighting over a female.

Overall then, species that have evolved by mental arms race amplifier selection will be radically different from species that have not so evolved. Nonetheless, they will not be abiological in their behavior; they will remain Darwinian organisms. On this model, one can explain almost all the seemingly Darwinian behavior invoked by the evolutionary psychologists, from sexual preferences to patterns of murder, while rejecting their attempts to force the analysis of human behavior into the narrow adaptationist framework that has been used for the analysis of social insects and other animals. At the same time, one can account for the spectacular flexibility of human behavior, particularly the ease with which millions of people can change their behavior over just two or three years. Thus accepting Darwinism as the foundation for human behavior does not require one to buy into the silky reductionism of evolutionary psychology. There is an alternative: immanent Darwinism.

THE ABSENCE OF SUBJECTIVE DARWINISM

A major problem facing this theory of generalized, open-ended, Darwinian calculation is that we still don't experience our mental processes as having Darwinian ends. It would be straightforward if humans rationally calculated the fitness contingencies of their actions, and then chose the course of action that would increase their Darwinian fitness. Some human behavior superficially fits this view, particularly human economic behavior, which often seems like rational self-aggrandizement. On the other hand, much human behavior appears not only irrationally self-destructive, but also heedless of fitness outcomes. Vasectomies are only one example of human behavior that is difficult to explain on the basis of rational Darwinism. Then there are such phenomena as substance abuse and smoking. Finally, there is the fundamental problem that, if most people calculate Darwinian plans of action, they certainly aren't aware of it introspectively. Net Darwinian fitness doesn't figure in the great lyric poems, or even in the treatises of political philosophers. The whole idea that we are busily planning our lives with fitness aforethought seems to falsify the record of human culture since our first records of that thought, from thousands of years ago. If it is something that we all do, but don't admit socially, then we are far better at keeping secrets than the contents of centuries of personal correspondence suggest. In total, it would be astonishing if a theory of human nature based on universal, self-conscious, Darwinian motivation should turn out to be correct.

Any solution to this problem other than the evolutionary psychology of genetically based behavior must involve a dynamic unconscious. Such a dynamic unconscious would be analogous to Freud's superego, in some ways. However, it would be driven by a Darwinian calculus, not some hydraulic or energetic imperative like Freud's. That is, the application of open-ended calculation to the human case requires the supposition of parts of the brain that we do not subjectively experience, but which orchestrate the functioning of the organism to Darwinian ends. In effect, such brain functions would profoundly regulate the specific calculations and functions of the rest of the nervous system according to ultimately Darwinian calculations. On this model, our subjective experiences and calculations would be like dogs on a leash, the leash held by a Darwinian master of whom we are not normally aware.

This type of theory patently invites incredulity. Among other things, it solves an empirical problem by the wholesale hypothesis of an "unseen mover." This is distasteful to any methodologist of science. In addition, the unseen mover is supposed to be of a type that would be almost impossible to observe directly, an unconscious brain function, or functions, that supervise other brain functions so as to foster Darwinian ends. Such an extravagant hypothesis, at first sight, makes evolutionary psychology attractive by contrast. But, as will be described shortly, there is considerable clinical evidence for the existence of the type of cryptic, Darwinian coordination just outlined.

However, for the time being a different point might be made. It is clearly possible to develop models for the evolution of human behavior that do not rest on the simple extension of principles from the study of animal behavior. In that sense, it is not a case of accepting evolutionary psychology or going back to behaviorism and other exploded social science theories. The theory of immanent Darwinism discussed here may not survive sustained scientific scrutiny, but it illustrates the possibilities for Darwinian theories fully appropriate to the special case of the human species.

In any case, to pursue this theoretical discussion further, a concrete fact about human social behavior must be introduced first: sociopathy.

SOCIOPATHY

The crucial group for testing the theory of unconscious Darwinism is called "sociopaths." This group has a distinctive psychological profile on testing, attaining high levels for "psychopathic deviate," "schizophrenia," and "hypomania" indices. It has been estimated that individuals like this commit about half of all crimes, even though they are only 1–2 percent of the population as a whole.[11] Not only is this group well known in criminology, it is also well known in psychiatry, where they are categorized in terms of a personality disorder variously known as sociopathy, psychopathy, and antisocial personality disorder. Here we call these individuals sociopaths, because this term has the most specific etymology.

The best portrait of the sociopath is that drawn by Hervey Cleckley in *The Mask of Sanity*.[12] Cleckley's description is of great value particularly because it is not oriented around criminological questions like crime rate, arrest rate, and the like. Instead, Cleckley describes what these people are like when they get under your skin. The single most devastating thing about sociopaths is that they can seem normal, intelligent, even charming on first acquaintance. Indeed, sociopaths are strikingly free of the many quirks that we look out for in the potentially dangerous. They don't necessarily have a fierce stare, they don't always smell bad, and they aren't usually taciturn. Rather, they can be loquacious and unaffected, virtually the model of open-minded generosity. As such, they are among the most effective con artists known to law enforcement. From a psychiatric standpoint, this group that is responsible for much of the crime and moral devastation of modern life nonetheless seems essentially normal. Sociopaths are not schizophrenic, intermittently psychotic, manic, depressive, suicidal, or neurotic, except when they are malingering. None of the major psychopathologies are part of their nature. Instead, many psychiatrists, Cleckley included, are often struck by how "well adjusted" sociopaths are, how free of the ticks that mar most of us in our presentation to the world. Sociopaths are not merely good hustlers, full of fake enthusiasms and friendliness. Their enthusiasms and concerns seem entirely genuine and convincing. For this reason, Cleckley refers to sociopaths as possessing "the mask of sanity."

But the life of a sociopath is radically disordered, whether or not they are incarcerated. They do not successfully pursue long-term careers. They do not remain faithful to spouses. They do not honor long-term promises of any kind. There is no ability to pursue a distant goal. Instead, they pursue impulsive objectives, frequently announcing that they have abandoned all their previous goals as out of keeping with their new understanding of themselves. Nothing is more common than pledges of reform made to family or partners, with a complete failure to follow the promised course of rehabilitation. Sociopaths burn out the hope of those who care for them.

While pursuing their erratic courses, sociopaths show little sign of inhibitions deriving from the needs or goals of others. They lack empathy, compassion, and mercy, except at the level of verbal protestation. This is not to say that they cannot behave generously or kindly, on occasion. Sociopaths do not read from a script marked "evil villain." Rather, they seem more heedless of everyday morality than actively opposed to it. But by normal standards, their conduct is despicable. They routinely lie, cheat, and steal. They assault and they murder. Prostitution is commonplace among sociopathic females. They abuse substances and behave obscenely in public. And in all such misbehaving, they typically end up in jail. It is estimated by various criminologists that sociopaths make up from 20 to 50 percent of the American convict population. Thus, sociopaths not only ruin the lives of those around them, they ruin their own lives as well.

This type of sociological description of sociopathy may seem unfamiliar, or even implausible. But the plausible reality of the sociopath has been well established within the medium of cinema. The sociopath's character has been brought to life in many movies of the *film noir* genre, from the performance of Joseph Cotten in *Shadow of a Doubt* to Linda Fiorentino in *The Last Seduction*. In *film noir*, we confront characters capable of effortless evil, without any hesitation, agony, or remorse. These are sociopaths. Sociopaths also figure prominently in the more recent genre of unsparing crime fiction and television, in which documentarian attention to psychological accuracy is combined with the directly shocking impact of rotting bodies and killers with no scruples or inhibitions. Collectively, these works provide a compelling portrait of the nihilism of the sociopathic existence.

The sociopath is a dramatically deviant type of individual. Indeed, so much so that their behavioral profiles are discontinuous with those of other individuals, even other criminals. This is like the pattern of a schizophrenic, in its deviance. But unlike schizophrenics, sociopaths are not overtly odd. Sociopaths can blend into society seamlessly, at least for a time. Sociopaths seem, in some sense, functional. They certainly are not pervasively impaired in brain function; they can fall well within the normal range on an IQ test, and some score above average. So why are they different? What accounts for the bizarre behavior of these individuals, relatively rare, but catastrophic in their impact on the rest of the world, as well as themselves?

Darwinian Theories of Sociopathy

Several authors, most notably Linda Mealey, have proposed that sociopathy constitutes an evolutionarily adaptive strategy.[13] In particular, it has been hypothesized that, while most individuals pursue rule-following reciprocating behavior, sociopaths are cheaters who defect on their obligations. On this model, so long as most other players follow the rules, the sociopath is supposed to enjoy a considerable payoff by ignoring the rules and taking whatever is needed to enhance his or her Darwinian fitness. Thus lying, cheating, stealing, and adultery are expected to be the essence of the sociopathic lifestyle, a putatively successful pattern of behavior within the human species. On this adaptive-strategy interpretation, they are expected to be as successful, in fitness terms, as normal law-abiding folks. Note that this kind of typological analysis of human behavior is closely analogous to the reasoning that E. O. Wilson and other behavioral zoologists have developed for the caste-like behavior of ants, bees, and other insects. The greater sophistication of Mealey, however, aligns her more with the evolutionary psychologists; she knows more psychology and criminology. In any case, she clearly falls within the intellectual tradition of sociobiology and its descendants or variants.

An alternative interpretation for sociopaths can be derived from immanent Darwinism.[14] If the preeminent feature of human nature is unconscious immanent Darwinism, then there should be occasional individuals that have failure of this adaptation. Thus, we know that sight is an adaptation in part because of the grave consequences of being blind. A test of unconscious immanent Darwinism is that there should be individuals that exhibit normal conscious function, but lack properly organized lives due to the absence of the unconscious Darwinian regulator. The psychiatric category of "sociopath" includes individuals that usually lack conscious mental deficiencies, such as retardation, neurosis, or psychosis, yet are unable to sustain successful families or careers. These individuals may therefore lack unconscious mental faculties that coordinate behavior in such a way as to foster enhanced fitness. An interesting feature of the behavior of sociopaths is that they are both cunning felons, in carrying out any given criminal act, and relatively easy to catch. They don't exhibit enough foresight to avoid even quite cumbersome efforts to track them down. On the interpretation of immanent Darwinism, sociopaths are "Darwinian idiots," even when they possess high IQs, defective humans that have lost their ability to mount any kind of Darwinian strategy.

Some cases of frontal lobe damage also seem to follow this pattern, most famously Phineas Gage.[15] Gage was a railway worker who had an accident in which a metal rod passed through the front of his brain. After his recovery, his behavior was dramatically transformed from that of a reliable employee to a belligerent and unemployable wanderer, though there is no sign that his IQ or basic mental abilities were affected. Both such brain lesion patients and sociopaths can be interpreted as individuals who are not following a cheating strategy, as such, but are devoid of any strategic facility at all. One of Cleckley's sentences describing sociopaths as a group resonates with this interpretation: "We are dealing here not with a complete man at all but with something that suggests a subtly constructed reflex machine." On this second interpretation, sociopaths have lost a critical function. They are pathological. As such, the expectation is that they are selected against—they should have lower Darwinian fitness.

It may be asked, how can we have sociopaths at all, if they do not follow some type of adaptive strategy? First, as mentioned, frontal lobe brain damage can produce sociopaths, or close approximations to them. Thus, sociopaths may be victims of perinatal and other developmental brain damage, rather than some type of genetic mutation. Second, they constitute a small fraction of society. The frequency of sociopaths is on the order of 1–2 percent, males and females combined. (Bipolar affective disorder has a frequency of the same magnitude, and it is a largely genetic disorder. As it causes at least a 25 percent death rate, it is unlikely to be beneficial.) Therefore, recurrent deleterious mutation, developmental accidents, and frontal lobe trauma may be all that is required to explain the persistence of the sociopathic syndrome. In any case, the decisive answer will come from estimates of the Darwinian fitness of sociopaths compared with normal individuals. If sociopaths are indeed well adapted, then they should have equal fitness to that of nonsociopaths. If they have reduced fitness, then they may be a strategy-null pathology. And the latter result would suggest the importance of unconscious mechanisms for the coordination of behavior to fitness-relevant ends. This would be one of the best pieces of evidence in favor of the theory of unconscious immanent Darwinism.

AN END TO INDOLENCE

To recap, it is far from the case that evolutionary biologists have generally sought to impose their views of the evolution of human behavior on the academic community, or the public at large. Most of those who have done so in recent times have been anthropologists and zoologists who are not primarily evolutionary biologists. Their actions have, however, forced the involvement of evolutionary biologists in questions concerning human evolution. Much of that initial involvement has been reactive, often hostile, toward those who would disturb the tranquillity of evolutionary biology.

Now there is the prospect of developing theories and experimental approaches to human evolution that are not merely awkward extensions of evolutionary research on animal behavior. Such theories can instead be based on the idea that human behavior may be coordinated to Darwinian ends by mechanisms dissimilar to those that shape the behavior of virtually all other animal species, immanent Darwinism. And there may be still other theoretical alternatives worth investigating. It is no longer the case that we have to chose between naive adaptationism and ideological posturing, where human nature is concerned.

10 SOCIETY

Ideology as Biology

EVOLUTIONARY THEORIES as presumptive as those of evolutionary psychology and immanent Darwinism define, or at least circumscribe, human nature. The human natures that these different theories entail in turn define, or circumscribe, the possibilities for human societies. To take an extreme case, the nature of sea anemones defines the kind of society that sea anemones can have, an extremely limited one. Biology must provide the ultimate constraints on society, even if these constraints are extremely lax for some species.

This intellectual issue isn't as odd or as novel as it might seem. There are long-standing traditions of economic theory and political philosophy in which assumptions about human nature are used as a starting point for analyses of human society, actual or prospective. Often the assumptions about human nature have little to do with any evolutionary understanding, when not actually fanciful. Letting anarchists, fascists, and utilitarians of no particular qualification dominate the field of discourse about human nature seems significantly irresponsible on the part of evolutionary biologists. Some sensible Darwinian model for human nature must underlie credible theories of human society. Unless one accepts some type of non-Darwinian theology of creation, we evolved by Darwinian mechanisms, according to Darwinian constraints. Karl Marx, for one, was willing to accept this starting point. But then so was Herbert Spencer. Here we will see how far it will take us in considering the main social science theories and political ideologies of the modern world. We will explore the biology behind ideology and social science.

Our starting point will be understandings of human behavior derived from a Darwinian standpoint. We have considered two fundamental alternatives when it comes to the Darwinian explanation of human behavior, evolutionary psychology and immanent Darwinian calculation. We have already seen the essential tension between these two views. This tension runs throughout our discussion of human societies.

HUMAN VALUES

One of the foundational questions for theories of human nature is the nature of human values. What do humans desire? What do they avoid? This is a question for which the two main Darwinian theories of human nature provide interesting answers.

Darwinian theories of human nature are agreed that the ultimate foundation for human values is Darwinian fitness. They also agree, however, that human behavior is not determined directly by this foundation. In the case of evolutionary psychology, fitness affects human behavior only indirectly, through the effect of genetic variation in behavior on the outcome of natural selection. Thus, the valuational calculus of human behavior is dependent on the outcome of genetic selection, for evolutionary psychology. For immanent Darwinism, Darwinian fitness also plays an indirect role, via the calculations of unconscious parts of the brain, particularly the frontal lobe. Biological evolution created the Darwinian machinery of the frontal lobe, but now genetic evolution plays little direct role in shaping human behavior. In its place, immanent Darwinian calculations take place in a bicameral mind. With either theory, the connection between fitness and human values is not expected to be extremely simple, and on the second theory it is expected to be quite indirect indeed.

This makes the prospects for using these different theories of value seemingly remote. Generally speaking, there is probably little likelihood of resolving the details of most human choices in terms of these evolutionary theories. But there may be one case where some illumination is possible. Value is a classic problem in the theory of economics.[1] It arises from the ostensibly simple question of how prices are determined in a market economy. It may provide something of an arena to evaluate, and contrast, the two evolutionary theories of human nature.

Modern theories of economic value mostly revolve around such arcana as "revealed preference." In these theories, the value of a commodity is simply the value that is made manifest by the choices of consumers, whatever the basis of those choices may be. This solution to the problem just pushes it back one step, to the minds of otherwise inscrutable consumers. In a sense, mainstream economics has given up on solving the problem of value.

The two main variants of the evolutionary biology of human behavior provide somewhat similar solutions to this problem of value. Evolutionary psychology presumes that human behavior is driven by a large collection of specific behavioral mechanisms, each the product of natural selection. These distinct mechanisms must operate by a particular set of genetically founded, though not necessarily environment-independent, valuations or goals. So, for example, mate choice must be driven by its own distinctive foci of evaluation, foci which will foster appropriate mating in most individuals, most of the time. These valuations might involve specific appraisals of such attributes of the prospective mate as their potential as a biological partner in fertilization, their ability to provide parental care to the offspring, and so on. Similarly, on evolutionary psychology, specific market commodities will be evaluated in terms of the degree to which they conform to these specific valuations. Thus, cars might be secondarily valued according to their inferred value specifically for mating, perhaps, where this valuation will be composed of specific domain-appropriate reactions. These may be positive or negative reactions to color, speed, style, and so on. The unknowing decisions of car designers will trigger these built-in reactions, producing the specific valuation of each car, and thus finally its price. Presumably, on this theory, car designers would be more successful if they knew more evolutionary psychology.

Theories of human behavior based on the idea of a general, at least partly unconscious, process of Darwinian calculation are in a somewhat similar position. On such theory, all product valuation must depend on a Darwinian calculus. But in this second type of evolutionary theory for human behavior, the valuation is specific to each individual at a particular time. There will be few built-in, well-defined, preferences that are already determined by genetic adaptation. Those producing products for markets will instead find that their customers exhibit baffling, sometimes transitory, valuations of their products, with variations between individuals that are often huge. On this theory, economists might as well leave the problem of value to revealed preferences. It will be hard for them to make much progress with the mercurial and unconscious evaluations of humans behaving as immanent Darwinian calculators.

Choosing between these two models of value comes down to the rapidity with which the prices in markets change according to demand. (Varying supply will change prices according to the relative scarcity of commodities, on either theory.) Basic consumer valuations must be much more stable on the evolutionary psychology model compared with their relative lack of stability on the immanent Darwinism model. Preferences that are genetically established must be both more general and more stable than preferences that reflect the calculations of many separate individuals. But whichever of the two evolutionary theories for human behavior is preferred, it is clear that both clarify the foundations of any theory of value in economics. Directly or indirectly, optimally or incorrectly, value must grow out of Darwinian mechanisms. This is one of the few points that all reasonable Darwinian theories of human society must agree on.

The Nature of the Market

Except at low levels of technology and low population densities, or complete determination of all exchange by a central authority, human societies will have some type of market made up of the exchanges, purchases, and other transactions of the individuals in any given locale. Most markets are riddled with imperfections of monopoly, restricted access, price-fixing, and miscellaneous government interventions. But they are all still markets, in which numerous individuals make their decisions as to production, consumption, and contractual obligation.

The two alternative evolutionary conceptions of human nature have some salience for our understanding of the workings of such economies. On the evolutionary psychology theory, the behavior of humans as producers, consumers, or intermediaries will be defined by specific behavioral mechanisms established by genetic adaptation. As such, these mechanisms should roughly define limits to the amount of change possible in any particular economy, locking individuals into patterns of consumption or production that could be highly inefficient, from the standpoint of the utilization of capital or indeed any other criterion. The genetic constraints of this type of evolutionarily defined economy should provide both stability of production and stability of consumption, and within this context prices might then fluctuate according to the market mechanisms that excite the interest of economists. Such an economy would be a predictable entity, amenable to the kind of macroeconomic analysis promoted by John Maynard Keynes.[2]

The alternative model is unconscious, open-ended, Darwinian calculation by an organism that has been selected specifically to be able to outwit other members of its species as well as a capricious environment, as described in chapters 8 and 9. In this case, behavioral adaptations will not usually be based on genetically determined mechanisms. Instead, creative unpredictability would be a hallmark. Such an organism would be a much more creative producer, and a much less predictable consumer, as it explored a wide range of routes to the enhancement of its fitness. As to the range and deviousness of its manipulations as an intermediary to commerce, there might be no limit. This Darwinian entity should be able to do any and all of the following: invade markets with completely novel products that eliminate extant products in the marketplace; switch rapidly from product to product as a consumer, according to new information about the utility of different products; and arrange complex deals involving groups of producers and savers, in order to bring products to markets.

The importance of the contrast between these two evolutionary scenarios for *Homo economicus* is the nature of the economy that results. On the evolutionary psychology view, the market can function as a durable clearinghouse, in which stable patterns of production meet stable patterns of consumption, the market having a minor role of adjusting supply and demand. On the second model, with open-ended, immanent, Darwinian calculation, the economy becomes a kind of melee, barely held together by institutional frameworks, with congeries of Darwinian fiends struggling for advantage. Such an economy would have stability only to the extent that technology remains at a low level. As technology advances, and there are more possibilities for the production of goods, the economy should become progressively more elaborate and less stable.

This contrast of consequences from the two basic evolutionary alternatives corresponds to a long-standing debate within economics in the twentieth century. The dominant neo-Keynesian and

monetarist schools have emphasized their ability to predict macro-
scopic behavior of economies, based on fixed relationships between
variables like the supply of money, the rate of inflation, the rate of
unemployment, and the rate of economic growth. The idea of such
stable features of economies accords well with the theories of evolutionary psychology.

An alternative, minority, approach within economics is that of
the neo-Austrians, such as Friedrich von Hayek, Ludwig von
Mises, and Joseph Schumpeter, who have emphasized the potential
of the market to change to such an extent that seemingly fixed relationships between macroeconomic variables are undermined.[3] Of
greatest importance for the present purposes is the way the neo-
Austrians construe the functioning of the market. In Schumpeter's
words, a market economy produces a "gale of destruction" in
which old ways of producing goods and services are continually
being destroyed by new inventions and practices. For neo-Austrians, the interesting and important aspect of the market economy is
the opportunity that it affords producers to bring better products
to market, qualitatively new products, which consumers can then
purchase. The macroeconomic tradition focuses more on the aggregate features of the economy, where such a focus is premised on a
less turbulent model of the economy. Thus, there is an unusually
direct correspondence between the two evolutionary models for
human behavior and two main types of economic theory.

THE POLITICAL STABILITY OF MARKET RULES

There is a long-standing tradition of Western philosophy,
ideology, and economics that treats the historical evolution of markets as if it were some kind of automatic process. This is as true of
the Whig visions of Adam Smith and his disciples as it is of Karl
Marx's historicist model of the progression of human societies
through feudal, capitalist, and socialist stages. This common belief
makes some important assumptions about the limits to human per-
ception and action.

On the model of evolutionary psychologists, it is perfectly reasonable to suppose that human societies could sleepwalk through
history, undergoing a succession of upward and progressive, or
perhaps regressive, transitions. On their model, fundamental ge-

netic constraints lock human behavior into patterns that it cannot escape. On their model, all the different forms of historicism and the like have a reasonable chance of being true.

On the second model, with immanent Darwinian calculation, no such stability can be expected for any historical process. To be concrete, we can illustrate this idea with the case of market intervention within a political system. As particular producers or consumers see their economic standing rise or fall, in relative or absolute terms, they can be expected to perceive this fluctuation. With open-ended calculation as a human capacity, it is only a matter of time before particular economic connections come to be perceived as the sources of difficulty for some economic agents. With political machinery adequate to the task, such as that provided by the modern Western state, it becomes conceivable for specific groups to ally together to protect their interests. These groups could then canvass the state, or even seize control, to get the state to intervene in the economy so as to give a particular desired outcome. Thus, manufacturers might lobby a government to exclude imports from countries that can produce the same goods for a lower price. Trade unions can compel governments to pass laws giving them "closed shops," so that all workers have to be members of a trade union. Consumers can lobby the government to ban particular products, subsidize the production of products, or provide them entirely free. With immanent Darwinism, human nature should be highly predisposed to bend, evade, or change any rule or procedure that impedes its progress toward reproduction. Humans will not continually react in the same ways, following some atavistic pattern produced by selection for foraging on the African savanna a million years ago.

From this, we can see that immanent Darwinism produces both great potential for market innovation and great potential for intervention into markets, intervention which will not usually foster the efficiency of the market system. In a word, there will be corruption. This might be corruption by an inner party elite, as in the Soviet Union or present-day China. Or it might be political competition vying for market spoils in a welfare state, like those of Western Europe. Or it might be the corruption of Third World nepotism, where members of the ruling family are given preferential treatment in the marketplace.

This is in any case just one example of the potential for humans to pervert governmental and economic institutions, granting the assumption of open-ended Darwinian calculation. A very general corollary of this evolutionary theory, then, is that no theory of history should work for long, once reasonable technological sophistication has been achieved. There should be too much creativity and corrosiveness for simple patterns of human social evolution to be long sustained. In effect, history becomes a tale full of sound and fury, signifying nothing, because there is no pattern to it. It's like the weather.[4]

TWO SCHEMES FOR ECONOMIC INEFFICIENCY

In a sense, both of the evolutionary models have bad news for economics. On the model from evolutionary psychology, humans are locked in on stereotyped patterns of behavior that have been established by genetic evolution arising from natural selection. Such organisms can be expected to lack the ability to respond to some types of environmental change. Thus, as producers, they will switch from product to product only with difficulty. As consumers, they will realize the relative value of different products only slowly. As intermediaries, they will perceive novel combinations of advantageous economic circumstances only rarely. On the positive side, this type of creature should "behave" well in a market setting, where corruption is concerned. Only rarely would this type of human corrode the basic fabric of a political economy.

The alternative model is perhaps worse news for economists. The creative fiend of immanent Darwinism has the potential to switch restlessly from occupation to occupation, product to product, inventing new and more obscure ways to evade laws and regulations. This gives some considerable capacity to invent new things and new processes, and thus provides markets with gales of destructive creation. Completely novel products should appear, and some of these might be taken up and spread through the economy with great rapidity, annihilating old industries and practices. But at the same time as this player in the economy supplies new and useful products, this player will also attempt to bend every rule, twist every monarch or politician, toward the interests of the player.

Thus, the economy can become bogged down in multiplayer grid-lock, in which orderly economic processes of any kind, capitalist or socialist, are subverted for the sake of the "interests" having greatest power. Such an economic system is like a dynamo with gravel thrown in, its motion slowed down or stopped by the friction of the struggle for advantage.

Both of these models, therefore, can explain the variegated fiasco that the modern global economy has become, in which all nation-states seem to contain reserves of productivity limited by perversities of market and political failure. Which model is a more compelling explanation is an open question.

HOBBES AND LEVIATHAN

The real world cares little for academic categories and conventions. The serious movers and shakers of every stripe often meet each other and appropriate each other's ideas. Themes from one area then show up in another, as poetry becomes politics becomes philosophy and then science. In the case of Thomas Hobbes, the great English philosopher, one such connection was made when he met Galileo. Like many thinking people, Hobbes saw that the Galilean system of unified mechanics offered a new way to think about all things. That is, Hobbes saw the universal scientific style of reasoning in Galileo's work. This style, of course, has since become one of the key foundations of the modern world. Hobbes wanted to create a "physics" of human social behavior, though neither he nor Galileo would have had a fully modern sense of these ideas. Hobbes's basic project was to imitate Galileo, but in a different sphere.[5] Some hundreds of theorists later, people are still trying to accomplish this objective, the evolutionary psychologists included.

Imitation may be flattery, but it is rarely very accurate. Hobbes took his inspiration from Galileo but produced something hardly as well reasoned as the mechanistic constructions of Galilean physics. Instead of building testable, dynamical models, Hobbes built an argument for the all-powerful Protector on earth, the Leviathan.

At the core of Hobbes's Leviathan is a fear of anarchy. The Reformation had set off a number of religious wars in Europe, some of which were appalling for their cruelty. Besieged cities suffered the plague, which was often deliberately triggered by the catapulting of diseased corpses over city walls. Babies were roasted on spits for public consumption. The inference that Hobbes and others drew from these events was that the state of nature must have been horrible: nasty, brutish, and short, in his famous words. Moreover, the narrow-minded European explorers of other continents were full of unflattering anecdotes about the supposed barbarity of the various aboriginal peoples they encountered. Not only were these peoples ignorant of Christianity, but they were often largely naked, disorderly, cannibalistic, and openly promiscuous. The conclusion of those like Hobbes was that man is bestial and loathsome when left to his own devices.

The problem then is how to save mankind from itself. The logical solution for the fearful is the creation of a civil power, the Leviathan, in the form of a Great Prince. This authority has to be given ultimate powers by the people in order that they may be saved from their own tendencies to evil and disorder. To Hobbes, this was the fundamental rationale for the state. This was also the root, or archetypal, pattern of authoritarian or conservative thinking, in the sense in which the term is used in Europe. Obviously this is not necessarily the conservatism of the United States, which is usually based on a belief in the merits of the original interpretation of the Constitution of the United States, a classically liberal document. If the terms authoritarian or conservative are disliked, one can substitute Tory or reactionary. The point is that this pattern of thought has been a self-sustaining current in Western thought. In addition, it has been an element in the thinking of other cultures as well. For example, many aspects of Confucian thought can be seen as an endorsement of orderly and order-preserving behavior in the face of temptations to behave in a selfishly destructive manner. Indeed, most traditional cultures have strong elements of what is called authoritarian thinking here.

But the biology that underlies authoritarian thinking deserves some attention. This model has the implicit assumption that humans are inherently inclined to behave in a fashion which is at once both destructive and selfish. On the authoritarian model, only investing a state with great powers will hold anarchy at bay. Thus, desperate societies turn to dictators whom they chose to invest with all civic power, examples ranging from Napoleon to Franco to Hitler. And the monarchies of Europe were often very explicitly derivative from the fear of anarchy and civil war.

The central flaw of authoritarian thinking is that it imagines that restrained behavior is a creation of the strong state. Nothing could be further from the truth. As presented in chapter 9, there is a human type who conforms to the myths of conservatism, the sociopath. Sociopaths are indeed reckless of convention, morality, law, propriety, or human decency, unless closely supervised. They kill children, their own parents, and hapless strangers, often for little or no motive. If sufficiently concupiscent, they will have sex with virtually anyone fitting their particular desires, in a wide range of combinations. So we can concede this much to the authoritarians. Their nightmarish notion of human nature has some existence, in this tiny minority. Police officers often seem to develop a Hobbesian model for human nature, which may arise from their frequent dealings with sociopaths. And police officers probably have more tendency to authoritarian politics than any other group. But sociopaths are a tiny group, 1–2 percent of the population. The vast majority of the species naturally obeys local laws and customs, even in the conspicuous absence of anything approaching an intrusive state.

The fundamental Darwinian justification for the rejection of authoritarian paranoia is that it is very unlikely that animals, humans included, would evolve so that most individuals lost all sense of hierarchy, property, and peaceful behavior. Retaliator and Bourgeois strategies are common to many animal species, from insects to birds to mammals, as discussed in chapter 3. They do not require any police, much less a police state, to be sustained. People likewise seem to develop ideas of marriage, fidelity, property, and self-restraint across all cultures. The best Darwinian strategy is not necessarily to behave like a rapacious seventeenth-century mercenary. Frequently, the best strategy, from a Darwinian standpoint, is to behave with restraint. And so most animal species are restrained in their social behavior, most of the time. It should also be noted that the evolutionary psychology and immanent Darwinism models are in agreement on this point. Both of them imply that authoritarian paranoias are fundamentally erroneous. The Hobbesian model of a "solitary, poor, nasty, brutish, and short" life without an all-powerful ruler is profoundly flawed.

CLASSICAL LIBERALISM TO LIBERTARIANISM

Darwinism was an outgrowth of the same intellectual movement that gave birth to economics and the liberal ideology, that movement being the British Enlightenment, particularly in its Scottish form. Even to say that Darwinism is a late offshoot of the Enlightenment as a whole is not an adequate characterization of Darwinism's roots. It is a scientific worldview that is British in origin, British and Anglo-American in its development, and to this day preeminently practiced in the United Kingdom and the United States, with lesser contributions from Canada and Australia. Karl Marx himself noted the Englishness of Darwin.[6] Darwin grew up reading the works of Adam Smith and other materials from the broad liberal tradition of which both his grandfathers were such exemplars.[7] Thus, he had in some part of his mind the ideas of progress, free trade, the breaking down of tradition, and so on. As a middle-aged man, in his overt political opinions, he remained true to this tradition in being opposed to slavery and generally rejecting the Tory side in British politics.

Thus, Darwinism can be seen as a natural development from the same British patterns of thought that gave rise to liberalism, in the classical sense. (This ideology is also known as "Whig" in Britain and "conservative" in contemporary America.) But the connections are still more extensive. Much of the specific content of Darwinism had elements derived from the thought of economists. The correspondence between Adam Smith's invisible hand of capitalism and the evolution of adaptation by natural selection is almost too obvious. The analogy between efficient firms and beneficial traits and their respective fates in the marketplace and the population is transparent. The better eliminate the inferior. This cruel-to-be-kind ethos lies at the core of both classical liberalism and Darwinism.

Unfortunately, this intellectual ancestry got Darwinism into trouble: "social Darwinism." Social Darwinism was not a creation of Darwin. Indeed, it really preceded Darwin, although the use of the term followed Darwin's death. As the term is normally used, social Darwinism means no more than the most hard-hearted form of classical liberalism, the doctrine of Ebenezer Scrooge. Classical liberalism had long taken a cool view of support for the poor and the disorderly. Like Dickensian villains, many of the classical liberals, such as Thomas Malthus, argued that charity only encouraged the poor to go on reproducing. In maintaining hard but fair competition, it was argued, the state would best serve the long-term interests of the poor. Thus, workhouses were made to be harsh, so that the poor would strive to leave them. All these ideas were in place before the *Origin of Species* was first published.

But Darwinism was a great intellectual achievement, and as with all such achievements, intellectuals and politicians attempted to hijack the theory to advance their own programs. And because Darwinism was partly classical British economic thinking applied to biology, it was only natural that late Victorian liberalism would exploit it first. (Later, as we have seen in chapter 7, fascism would make its attempt to exploit Darwinism during the heyday of eugenics.) The most notable of the Darwinian intellectuals outside of biology was Herbert Spencer. Spencer was the editor of *The Economist*, a periodical that carries the banner of the classical liberal cause to this very day. He also had the lack of taste to brush aside the love of the great George Eliot (a.k.a. Marian Evans). He wrote endless tomes integrating all manner of knowledge into elephantine syntheses of little clear import. Finally, he saddled Darwinism with the unfortunate, and actually misleading, phrase, "survival of the fittest." From people like Spencer, lesser figures absorbed the message that they could generalize Darwinism to any phenomenon or issue that they liked. And so Darwin was swept into service as another element justifying the strictures of classical liberalism. After all, the story went, if Darwin has shown that competition in nature leads to the survival of the fittest, and the improvement of forms of life, then surely we should use this same doctrine in our governmental policies concerning the poor, minimum wages, pensions, and so on.[8]

The extensive depressions between the two world wars essentially destroyed classical liberalism as a dominant ideology. (Even Friedrich Hayek, the neo-Austrian Nobel Laureate in Economics and the leading advocate of a reborn liberalism, was a socialist at first.) The good thing about this situation was that Darwinism was let off the hook. The leading evolutionary biologists of the 1930s were free to be communists, fascists, etc. J.B.S. Haldane was Stalinist. Other evolutionary biologists, like R. A. Fisher, were Tories and eugenicists. By that time, there was no longer a strong connection between Darwinism and classical liberalism.

But things would start to heat up again in the late 1960s and early 1970s. The "sixties" counterculture was mostly unsuited to good old-fashioned Marxism-Leninism, and even Trotskyism and Maoism were a bit too authoritarian for some of the radicals of the time. Bakunin and Kropot, among other anarchists, enjoyed more favor on college campuses in the sixties than they had since 1917. Not for all such anarchists, but for some, libertarian ideology proved to be a siren song. Perhaps the most eloquent of libertarian manifestos was Robert Nozick's *Anarchy, State, and Utopia*.[9] Nozick argued that a type of minimal state could develop by consent, starting from an initial condition of anarchy. Moreover, he argued against the extension of this state beyond the minimal elements of police, judicial system, and military. Thus, he rejected pensions, public works, welfare, and so on. This was essentially a bridge back to classical liberalism. And most important, for our present purposes, it was a treatise that relied heavily on notions of human behavior in a state of nature. Human nature again was a pivotal issue.

So long as the accepted theory of human nature is non-Darwinian, then of course libertarian ideologues can hypothesize a human nature that would fit their ideology. Every other ideology has done so, why not them? Admitting the constraint that human nature has to conform at least to the possible, as conceived in Darwinism, then we need to consider the implications of the two basic Darwinian models for human nature. Evolutionary psychology has the comforting feature for ideology of supposing that human behavior is constrained in specific ways. Thus, it is at least conceivable, assuming such theory, that some type of libertarian state could be configured so that citizens could go about their business with minimal interference inflicted on their fellow citizens. In effect, the correct ideology would then be one that best exploited both the "dumb" and the intelligent aspects of an evolutionarily well-structured, specifically human, and genetically stable behavioral repertoire.

No such assurance exists on the alternative Darwinian theory. If humans make open-ended calculations of Darwinian contingencies then the same basic problem of political and economic corruption described above comes back to haunt any reborn liberalism with a vengeance. Instead of resting content with some type of automatic market-based allocations of goods, individuals can form factions to corrupt the market and polity, not just on the small scale of market imperfections, like oligopolies, but by seeking wholesale interventions by the government. A libertarian society would be torn apart by factions. And one interpretation of American politics since 1830 is that a fairly libertarian constitution was shredded in the service of first one interest and then another. This is at least one example, admittedly not perfect, of the essential instability of libertarian government. It suggests that all ideological constitutions will be shredded in the course of politics, both revolutionary politics and retail politics. Thus, the problem of market stability has its counterpart in the problem of political stability in the face of humans that are immanent Darwinian calculators.

FROM ROUSSEAU TO ORWELL

There are few myths more alluring than the one famously spread by Jean-Jacques Rousseau, that humans are naturally good and kind.[10] This is almost the religion of those anthropologists who seem to be determined to become the entrée for the cannibal's next meal. It is also implicit or explicit in a large amount of Romantic Victorian literature. The Noble Savage is a good-feeling myth. It says that the natural human is a good human. The natural is good, the civilized bad. This idea has burrowed deep within the modern psyche. It is implicit in our child-rearing practices and our legal treatment of juvenile offenders, the imbecile, and the insane. Innocence is not just a presumption; it is a determined conclusion.

Since powerful and affluent adults in modern societies are not supposed to conform to this model, some excuse is required to preserve the underlying hypothesis of nobility. That model is usually alienation, oppression, or corruption by a civilized society that is supposedly organized in such a way as to thwart the essential goodness of mankind. Note that this model is almost the precise inverse of the paranoid Hobbesian model. In psychiatric terms, instead of being paranoid, Rousseau's vision is grandiose. Some suppressed tendency to universal benignity is assumed, a tendency that requires the destruction of the established order to enable the transformation of society into a new and congenial form. This is the ideology of revolution.

This style of thinking is extremely widespread in the modern world, mutating into a variety of forms. It has been a staple of twentieth-century journalism that governments, universities, and private corporations are malignant organizational systems ("the System" sometimes) that crush the human spirit. The dramatic leitmotif of the journalistic piece is then the crusade of the journalists, or their subjects, against this violation of ordinary humanity or decency. But surely the cruelty of the civil servant toward his or her victims is as much a manifestation of ordinary humanity as anything else? Sometimes the incoherence of thought is amazing, as in the younger George Orwell, who could stingingly sketch the deficiencies of the ordinary people that he met, in books like *Down and Out in Paris and London* or *The Road to Wigan Pier,* and yet suppose that these people would be radically ennobled by a different political organization of society.[11]

Overall, this basic coupling of a hypothetical original nobility and its subsequent Fall is one of the most common, elementary, ideological assumptions of modernity. (In a sense, it is as basic as Adam and Eve retold in modern dress.) And it patently makes strong assumptions about the basic features of human nature. While not all moderns subscribe to revolutionary politics in the waning years of the twentieth century, it was once the dominant creed among young people of every age during the sixties. The most elementary feature of this creed was hostility to the market system which, after the depressions of the interwar years, was seen as an archetypal example of the ways in which conventional political and economic systems ruin the lives of their citizens. Instead of this barbaric arrangement of free markets and democracy, socialist government would take over all the major assets of the country and deploy them only in the interests of all. Market pressures and all their attendant irrationalities or evils would be banished.

Let us take a step back and consider these issues from a Darwinian perspective. Socialism could be compatible with the evolutionary psychology system. Central to this supposition is that the evolutionary psychology model is basically an animal one. It hypothesizes that humans will conform to specific optimal behavior patterns that are genetically based. If this is true, then there should be ways to manipulate each of these behavioral patterns so as to maximize benign human behavior. In other words, evolutionary psychology leaves humans well suited to control by the "social engineer," the type of behavior controller who is implicit in all totalitarian revolutionary systems, and very explicit in such works as Orwell's *Nineteen Eighty-four* or Terry Gilliam's *Brazil.*

On the alternative model, immanent Darwinian calculation, socialism would have no more success than any other ideology. One of the basic problems posed by immanent Darwinism for such ideology is that the technocrats or commissars of the socialist state would be tracking a moving problem. Whichever policies they attempt to implement, given an initial appraisal of the behavior of their citizens, those citizens are then likely to modify their behavior to exploit that policy to their several Darwinian ends. And since the many separate calculations of all the individuals in the socialist society would endlessly find new "nooks" or "hooks" in the edifice of the state's policy, the overall structure would always be lurching forward, blindly, seeking a stable order that is not available. Stagnation, as exhibited by the Soviet Union during most of its existence, or disintegration, as exhibited by the Soviet empire around 1990, would seem inevitable. Humans behaving with their predicted spontaneity and creativity would be corrosive for any socialist state, however benign. On the hypothesis of immanent Darwinian calculation, there can be no peace for the social engineers.

THE END OF IDEOLOGY?

In 1960 Daniel Bell published a book that set off considerable debate and soul-searching, *The End of Ideology*.[12] The irony of his thesis was that the world was about to go through a generation of acute ideological conflict, ending only with the collapse of the Soviet Union and the conversion of China to capitalism. But now almost the entire world runs on some variant of capitalism, however corrupt or disguised, excepting only marginal countries like North Korea, which have death-watch status. For this reason, we hear news about "the end of history," in Fukuyama's phrase,[13] which echoes Bell.

However, ideology is something that is always being reborn, or at least repackaged. Whatever the interpretation, whatever the motive, the feasibility of any ideology rests in part on the fundamental nature of humans. On no reasonable evolutionary model for human behavior is authoritarian (a.k.a. reactionary, conservative, etc.) ideology well justified. Good human behavior does not require a police state. As to the more utopian ideologies of libertarianism and socialism, their viability depends on which particular evolutionary theory of human nature is preferred. On evolutionary psychology, it should be possible for social engineers to exploit the genetically determined aspects of human behavior to build some kind of viable capitalist or socialist state, despite the fact that no one has yet succeeded in doing so.

But on the alternative model for human nature, with open-ended Darwinian calculators, any ideology seems likely to fail. Such creatures would quickly become vermin, if not termites, within the edifice of the ideological state. In this case, in the words of Sting, "There's no political solution / To our troubled evolution."[14]

11 *RELIGION*

The Spectre Haunting

RELIGION AND EVOLUTION seem to be linked together in the public eye, like few other issues. It is interesting to ask how this came to be, and what is its significance.

Western civilization up until the time of Isaac Newton had a number of simple, powerful, and widely accepted beliefs which brought together almost all educated members of society. These were as follows. God had created the material world according to a divine plan. Men were creatures of spirit and flesh, only the latter being mortal. The world had been created to last only a short period before Final Judgment. In any case, the spiritual realm was the one of real importance, and the source of all order on earth. Accordingly, all disputes about causes and reasons were ultimately resolvable only in terms of theological arguments.

In this, Western civilization was unremarkable. All premodern civilizations had theologies, in which beings or forces of great power give rise to all apparent order on this earth. Such forces or beings were usually conceived of rationally, in order to make sense of seasons, fortunes in war, and so on. Before science, learning was dominated by religion, and men of learning were almost invariably priests. Like ideologies, religions are remarkably immune to experiment and evidence. If one is persistent in producing evidence or arguments which contradict an article of faith, one will be persecuted as a heretic, rather than praised for one's discovery. Religion is ultimately concerned with authority and faith, not doubt and knowledge. As such, it provides one of the key elements that bind the components of civilizations together, stabilizing the values of their societies.

Religious Significance of Physics

There is a popular myth that the rise of modern physics marked the end of the dominance of Christian theology over European thought. The sentencing of Galileo to death for his astronomical work has been taken to reflect the attempt of the Roman Catholic Church to save itself from encroaching science. In fact, Galileo was a pious Christian, who chiefly ran afoul of Aristotelian scholars who were exploiting the power of the Inquisition to destroy Galileo's empirical approach to physics.[1]

The man who embodied the true relationship between physics and Christianity, as it existed in his day, was Isaac Newton. Newton was the man who systematized mechanics, developing the basic conceptual tools of theoretical physics and fashioning the first mathematical cosmology. With such achievements, he held all of educated Europe spellbound. But he was in no way an enemy of religion. He was profoundly dedicated to theological ways of thinking, and avidly opposed atheists. His view was that the mathematical order shown by his work was merely a revelation of God's design, not evidence for a materialistic universe which could do without God. Working behind the scenes in the Royal Society, Newton actively fostered the dissemination of astrotheology, a school of thought devoted to presenting God as a divine geometer, responsible for the precision of the mathematical relationships which Newton had discovered underlying the orbits of the planets.[2]

Religious Significance of Darwinism

Charles Darwin's ideas convinced many that God had not created all living things, and that instead the genesis of the living order was to be explained in terms of blind, material causation. Most importantly for the general public, the origin of our species could be explained in materialistic terms. These ideas led to many crises of faith, and the vilification of Darwin by some of the devout. Unlike Newton, Darwin was set against the entire established order of Western Christendom, since he himself became an atheist. And Darwin's work led to the eventual unseating of Christianity from the center of Western scientific thought.

Perhaps the point which was crucial to the great importance of Darwinism for "scientific theology" was as follows. Life is both well organized and diverse. The notion that simple physical forces could have produced it simply shocked the rational mind before Darwin. Naturally, atheistic thinkers could develop simple theories about life based on remarkably crude physical ideas, and many did before the twentieth century. But those who knew more biology were staggered by the sheer improbability of such forces producing the range of life, from whales to birds of paradise, to say nothing of the immeasurable diversity of invertebrate animals and plant life. In particular, the exquisite suitedness of many components of the body to sustaining life was taken as evidence of the most careful, omnipotent design.

Few people expressed this better than William Paley, who argued that if we find a watch upon the heath, its intricate workings suggest that a powerful intelligence designed it. Likewise, the marvelous contrivances of eyes, ears, and webbed feet all suggest creation by a powerful intelligence, evidently God.[3] This whole style of thought culminated in the Bridgewater Treatises (1833–36), consisting of twelve volumes composed by eight authors, devoted to evidences of God's providence in the works of creation.

Instead of a beneficent creation, Darwin made a purely material genesis of life plausible. And not only life in general, but humans in particular, the species that religious cosmologies were often specially concerned with. This then marked the point from which science could simply abandon any religious latticework, and set about explaining the known universe on its own terms. And seeing that they could, many scientists, though not all, did.

THE CREATIONIST BACKLASH

A reaction wasn't long in coming, and has been long sustained. This backlash is now better known as creationism. Creationism is an intellectually articulate movement with some brilliant advocates. It is not a mob of illiterate malcontents. They wage their religious war with determination and subtlety.[4] And their target is evolutionary biology. For this reason, evolutionary biologists usually fight creationists on their own, unaided by scientists from other disciplines. But physicists and chemists have a big stake in the struggle with creationism too. If an omnipotent creature can intervene in events in the known universe, then there is no reason to suppose that any law of science will necessarily remain constant. Indeed, to the extent to which the processes under scientific study are subject to divine intervention, they become unsuited to further scientific study at all.

This can be made concrete. Let us suppose that a particular theory of superconductivity requires that a given alloy have a particular electrical resistance at 10° K. If that resistance is being measured in a laboratory at 3:15 P.M. on a Tuesday and the result is not what was expected by the theory, then a creationist scientist could say, simply, that God must have intervened to change the alloy so as to get the observed result. The alternative, which a noncreationist would be forced to embrace, is to admit that the original theory is wrong.

This is the essential clash between science and nonscience. With the latter, there is always room to move about to avoid accepting any evidence indicating that you, or your favored ideas, are wrong. This is what trial lawyers, campaigning politicians, and guilty five-year-olds do. Facing mistakes and moving on to change your ideas are central to science. All scientists would have the cogency of their work undermined by a successful creationism, a creationism in control of the schools, the universities, and the government funding agencies. To the extent that evolutionary biologists defend scientific turf from creationists, they are defending all of science.

SCIENCE, THE UNIVERSE, AND RELIGION

A simple reading of the relationship between Darwinism and religion might lead one to the conviction that Darwinism has been a long-standing enemy of religion. In some respects, this is true, but not in all. An often lost, but still important, distinction in discussions of the relationship between science and religion is the confusion of the pertinence of religion for the material organization of the universe with the pertinence of religion for the human experience of life. This is a natural confusion, because most religions, from Taoism to Christianity, are based solidly on this conflation of microcosm and macrocosm. However, there is no *a priori* necessity for it. Here we will consider these two facets of religion separately.

First, we review the question of science and religion in the explanation of the material universe. As we have already discussed, Darwin undermined divine creation as an explanation of the biological world by supplying an intellectually attractive, impersonal mechanism for the generation of adapted and diverse forms of life: evolution by natural selection. He could even account for numerous details and imperfections of life that were essentially inexplicable using the older creationist biology.

By contrast, continuing on from Newton, there is a long-standing tradition in physics that is essentially theological. While Laplace "had no need of that hypothesis," many physicists have regarded their work from a relatively theological perspective. Einstein was fond of saying that God does not play dice with the universe, when arguing against some of the tendencies in quantum mechanics that he regarded as perverse. Contemporary physicists persist in this vein, referring to their research as revealing God's Mind and the like. This pseudo-theological aspect of physics has attracted relatively little attention, perhaps because it is essentially Deist. That is, the God of physics is the God who sets the machinery of the universe in motion, but does not cause the face of a particular saint to appear in bread mold. It is a removed God, a distant God. Not a God who requires worship. Instead, a God who can barely be glimpsed until you have earned a Ph.D. in physics. Sometimes this God of physics seems to be little more than the universe itself.

There is nothing in the length and breadth of science that really requires this invocation of God to explain how the universe works. This is not to say that there aren't "holes" in science. There always will be. Many natural scientists discover these holes with relish and pursue them with determination. It is out of such holes that major improvements are made to science. And plugging these holes quickly and expeditiously by invoking God is entirely false to the spirit of the modern scientific enterprise. When physicists and other natural scientists resort to such stratagems, either they are advancing the cause of religion disingenuously or they are stupidly undermining the integrity of the scientific process. Overall, natural scientists should encourage the physicists to give up their rather precious allusions to, or invocations of, some type of transcendent God, unless these particular physicists want to come clean and confess to being theists. Each of us, after all, is entitled to our own beliefs.

DARWINISM AND THE VARIETIES OF RELIGIOUS
EXPERIENCE

Now we turn to the second aspect of the relationship between science and theology, the subjective experience of "the transcendent." "Theological science," particularly that of physics,
leaves aside the intense and commonplace experience of some type
of supernatural realm. This might be the experience of God during
evening prayer, or the intoxication of a sensual evil presence during
a voodoo rite. Experiences like these are so widespread, so strongly
proclaimed, that there must be something to them.[5]

The question is, what is that something? A simple answer offered
by the various devotees of one or another sect is that their favored
deity or deities is the source of their experience. The differing opinions of these different sects can then be accounted for in one of two
ways. The conventional way is to assert that the spiritual experiences other than those of your own denomination are somehow erroneous or perverted, possibly by an evil force. Less conventionally,
we might suppose that all these different deities exist, and different
religious or spiritual movements just "tune in" on different wave-
lengths, those occupied by their favorite deities. This last theory
will be understandable if one has lived in a big American city from
the ability of different radio stations to fix on a particular market
segment. The radio stations, on this model, are gods, while we, as
their listeners, are thus their worshippers.

Alternatively, we could suppose instead that all these fervent be-
lievers are fundamentally misled. Instead of connecting with a
deity during their "experiences of the Other," they are undergoing
some entirely prosaic process of human psychology. Two basic
models might then explain the "spiritual" in terms of basic psychology. The first is gullibility, hucksterism, and their infelicitous combination. On this model, the religious are rubes, and some people
supply them with a faith that leads to a net transfer of funds from
the flock to the shepherd. At least some examples of the religious
experience must come from this social dynamic, particularly that
of televangelism. But many religious experiences seem to arise
without any kind of social encouragement. And there is an additional problem posed by the success of the television evangelist.
What do they have to sell that people are willing to buy? There

must be some basic type of pleasure to be had from listening to hours of pious protestation. Hypnotism by television is not a sufficient explanation. Therefore, despite the great appeal of the classic "fools and villains" model for the religious experience, in the end it has to be regarded as inadequate.[6]

A second basic model is the idea of a dynamic unconscious, discussed in chapter 9. Freud and his disciples have long argued for the existence of a dynamic unconscious, although their model was more like plumbing than a modern notion of what a large part of the brain might do. The Freudian notion of the unconscious was virtually hydraulic, psychic energy flowing around, accumulating, and being dissipated. A modern conception of an unconscious would be much more concerned with the processing of information. Here we have already discussed the idea that humans have an extensive Darwinian unconscious which orchestrates our behavior to Darwinian ends. Our interpretation of sociopaths in particular is that they lack a functioning Darwinian unconscious, and thus their long-term behavior is incoherent.

Note that this hypothesis proposes the existence of a major entity functioning with long-term calculation within the same brain as that in which our conscious minds are embodied. This is a bicameral mind, somewhat like that which Julian Jaynes has suggested characterized premodern humans.[7] In the present model, as in Jaynes's analysis, the existence of the conscious mind together with a second unconscious "mind" naturally gives rise to some type of experience of the Other. On this model, the "Other" is the Darwinian unconscious. Its care and solicitation with respect to our fate is the care and solicitation incorporated in the ramshackle Transcendent/Immanent Gods of Christianity and other religions. Recall the metaphor of the dog and its master for the relationship between the conscious mind and the unconscious Darwinian regulator. The experience of this relationship, even if indirect, must be one of the most fundamental in human life. It may be the source of the religious visions of the psychotic, in that psychosis may break down the separation of the two cameral components of the mind in some cases. Stress, during emotional desolation or physical starvation, may have the same effect as psychosis, since the brain is just an organic computer, and its organizational integrity may be compromised by biochemical disruption. Even without stress or psychosis, normal people may experience an awareness of the looming involvement of normally undetected parts of the brain. Such experiences, both extreme and everyday, may be the basis of what is commonly thought of as religious experience.

This model may seem perverse to some Darwinians. It says that religious experiences are in some sense real. Not real in the sense of the "loaves and fishes" miracle, the "water into wine" miracle, or any other miracle of that kind. It is not supposed that any of our causal notions of physical reality can be violated. Rather, the proposal is that religious and other spiritual experiences involve our conscious mind's contact with another major element of our brain's functioning. In this limited sense, this theory is a kind of rapprochement between Darwinism and religion. In one sense, that of explaining human behavior, Darwinians need to take religion seriously. In another sense, that of religion making its way in the modern world of science and technology, the religious may obtain some relief from a model of religious experience that might be reconcilable with science. On the theory of immanent Darwinism, religious experiences may have some foundation in underlying reality, rather than being primitive superstitions in more elaborate dress.

Conclusion

AT THE END of it all, is it better to have known about Darwinism? Would mankind have been better off thinking that we all came from Adam and Eve, or a bear that mated with itself? Does Darwinism help our moral character, or hurt us? Is it any easier to find peace between religious sects, now that no one can reasonably suppose that religion is the best guide to our biology or our origins?

The case for Darwinism cannot be based on any edification that is supposed to come from its truths. Through eugenics, Darwinism was a bad influence on Nazism, one of the greatest killers in world history. Darwinism probably contributed to the upsurge of racism in the latter part of the nineteenth century, and thus it helped foment twentieth-century racism generally. Darwinism was also used to exacerbate the neglect of the poor in the nineteenth century. All things considered, Darwinism has had many regrettable, and sometimes actually vicious, effects on the social climate of the modern world. Modern Darwinism does not offer any guarantee of unending progress. It is understandable that so many hate Darwin and Darwinism. It is often a bitter burden to live with Darwinism and its implications. Unlike so many doctrines, religions, and ideologies, it certainly isn't intellectual opium. No one can make a case for Darwinism based on moral hygiene.

Almost the opposite is the case where the practical benefits of Darwinism are concerned. Darwinian thought is essential to livestock breeding, agronomy, and the like. Modern agriculture depends on Darwinism as one of its most important foundation stones. We are just beginning to see the use of Darwinian methodologies in medicine, genetic engineering, and cognate fields. But more such applications are sure to come. The fabric of modern civilization depends on Darwinism as one of its tent poles. Much would collapse without Darwinism, even though its contributions are not apparent at the surface of contemporary life.

Despite controversy, obfuscation, and a lot of bad temper, Darwinism is striving to help us understand human nature. This is perhaps the hardest thing of all, because of all the baggage that Darwinism carries around, and the appropriate reluctance of many to let us forget that baggage. Nonetheless, there are some faint signs that progress is being made with respect to the Darwinian self-examination of humankind.

The last thing is the most important of all. Darwinism has been a bad thing for social peace, and a good thing for several material endeavors, but for science itself it has played a key role in binding biology to the physical sciences. Without Darwinism, biological science would need one or more deities to explain the marvelous contrivances of life. Physics and chemistry alone are not enough. And so without Darwinism science would necessarily remain theistic, in whole or in part. This makes science without Darwinism an essential goal for those who want science to be subjugated to religious ministries. For those who believe that scientific truth is our best guide to material truth, even if not an infallible guide, Darwinism is a powerful ally. For those who seek more than the endless details of molecular biology, Darwin's Spectre is Virgil in a moronic Inferno of biological facts. And for those who would know and understand the long story of life on Earth, Darwinism is the great searchlight in the darkness. For the modern world to go on without Darwin's Spectre would be to lose our way in a twilight of the mind.

Bibliographic Material and Notes

I HAVE TRIED to preserve the text of the book in a readable form, rather than adhere to scholarly citation. However, there is a vast amount of information about Darwinism that is relevant to this work, and some of it is well worth reading. This bibliographic essay provides information about two kinds of reference.

The first kind are the general references that have been used by me in writing this book, and which would probably be useful to a reader wanting more background. These references are given at the start of the sections on the individual Parts of the book, together with some comments on their possible value. These comments are subjective, and should not be taken as scholarly evaluations. Another Darwinist could well have a different set of comments.

The second kind of reference supplied here are notes giving information about specific sources for assertions or quotations. These are grouped under chapter headings. I occasionally indicate whether such a source might be worth pursuing further. However, this is by no means an exhaustive list of all possible citations of sources. Academic completeness is not the primary motive in the notes.

General References and Readings for Part One

There are numerous marvelous books about Darwin and about Darwinism. Perhaps more than any other area of biology, Darwinism attracts excellent writers.

With respect to Darwin himself, there are many notable books. One place to begin is Darwin's autobiography, which has been published conveniently along with T. H. Huxley's (*Charles Darwin, Thomas Henry Huxley, Autobiographies*, ed. by G. de Beer, 1983, Oxford University Press, Oxford). There are a number of Darwin biographies, the best of which, in my opinion, is that of Janet Browne (*Charles Darwin, Voyaging*, 1995, A. A. Knopf, New York), although it remains incomplete. (A second volume is expected.) Browne's book gives a stronger flavor of Darwin than anything else I have read. Two other recent Darwin biographies are *Charles Darwin, A New Life* (J. Bowlby, 1990, W. W. Norton, New York) and *Darwin, The Life of a Tormented Evolutionist* (A. Desmond and J. Moore, 1991, Warner, New York). Bowlby tends to be dry, while Desmond and Moore are somewhat speculative psychologically. Another book of interest is Irving Stone's *The Origin, A Biographical Novel of Charles Darwin* (1980, Doubleday, Garden City, NY). Stone is honest about the fictional nature of his account, unlike some biographers, and his narrative powers are considerable. However, Browne still manages to give what seems to me greater verisimilitude to her portrait of the man.

Distinct from Darwin himself is the historical career of Darwinism, as a scientific movement. An accessible, if somewhat idiosyncratic, beginning is provided by Loren C. Eiseley's *Darwin's Century: Evolution and the Men Who Discovered It* (1958, Doubleday, Garden City, NY). Gertrude Himmelfarb's *Darwin and the Darwinian Revolution* (1962, 2d ed., W. W. Norton, New York) is that rarity, a true classic that is eminently readable. Also accessible is Ronald W. Clark's very professional *The Survival of Charles Darwin: A Biography of a Man and an Idea* (1984, Random House, New York). Perhaps the gem in this field is William B. Provine's *Origins of Theoretical Population Genetics* (1971, University of Chicago Press, Chicago). Despite its title, Provine's short book is a lucid and penetrating account of the development of evolutionary thought from the publication of Darwin's *Origin* up to the 1930s, by which time evolutionary genetics had solid foundations. A book that ranges from classical thought to the latest trends in evolutionary research is *Darwinism Evolving: System Dynamics and the Genealogy of Natural Selection* (D. J. Depew and B. H. Weber, 1995, MIT Press, Cambridge, MA); again, the forbidding title is somewhat misleading. The book is actually a fairly general, and detailed, intellectual history of Darwinism. Among Michael Ruse's many books, *The Darwinian Revolution: Science Red in Tooth and Claw* (1979, University of Chicago Press, Chicago) concentrates on the nineteenth century, and offers a well-balanced perspective. Ernst Mayr's *The Growth of Biological Thought, Diversity, Evolution, and Inheritance* (1982, Belknap Press, Cambridge, MA) is unusual in being written "from the inside" by an accomplished evolutionary biologist. It is a massive book, easily earning the sobriquet magisterial. Something of a distillation is provided by Mayr's *One Long Argument: Charles Darwin and the Genesis of Modern Evolutionary Thought* (1991, Harvard University Press, Cambridge, MA).

To this point, we have been discussing historical books that are not primarily scientific evaluation. To turn to the real scientific arguments, the starting point has to be Charles Darwin's *On the Origin of Species by Means of Natural Selection, or The Preservation of Favoured Races in the Struggle for Life* (originally published in 1859 by John Murray, London, but now available in many editions). This is perhaps the greatest original scientific book ever written, and it is surprisingly good in terms of style, as well. The reader should be warned, however, that few of Darwin's other books come close to the standard set by the *Origin*. Among these others, I recommend his *Descent of Man, and Selection in Relation to Sex* (1871, John Murray, London).

There is a group of outstanding scientific books by the leaders of twentieth-century evolutionary thought. Most of these books are out-of-date, but they can still generate a great deal of excitement for the prepared reader. A partial listing would have to include the following: R. A. Fisher, *The Genetical Theory of Natural Selection* (1930, Oxford University Press, Oxford); J.B.S. Haldane, *The Causes of Evolution* (1932, Longmans, London); Th. Dobzhansky, *Genetics and the Origin of Species* (1937, Columbia University Press, New York); and E. Maryr, *Systematics and the Origin of Species* (1944, Columbia University Press, New York). Perhaps the dreadnought in this class of books is Sewall Wright's *Evolution and the Genetics of Natural Population* (in 4 vols., 1968–78, University of Chicago Press, Chicago). It is the most up-to-date. It is also the most difficult for the general reader. For the general reader, I would be inclined to recommend the books by Haldane and Dobzhansky first.

Finally, we come to the many academic textbooks that give evolutionary biology in prechewed, predigested form. While such books are often quite stale, of necessity, they can clarify scientific questions to a greater degree than more topical books, like the present one, or more specialized books, like those written by the pioneers of the field. Among a sea of such books, the following are my personal favorites, starting with the more elementary and proceeding to the more advanced. John Maynard Smith's *The Theory of Evolution* (3d ed., 1975, Penguin Books, London) has the virtues of the author's extremely fine mind and his crystal-clear prose. An oddity for any scientific field is a popular account of a research project that leaves some of the details in, which is what Jonathan Weiner does in *The Beak of the Finch: A Story of Evolution in Our Time* (1994, A. A. Knopf, New York).

George C. Williams's *Adaptation and Natural Selection* (1966, Princeton University Press, Princeton, NJ) is a virtuoso workout of verbal Darwinian reasoning. Discursive and wide-ranging is Douglas J. Futuyma's *Evolutionary Biology* (2d ed., 1988, Sinauer, Sunderland, MA), though mathematics does creep in. A more conventional textbook is provided by Mark Ridley's *Evolution* (2d ed., 1996, Blackwell Science, Cambridge, MA). Among the advanced textbooks, D. L. Hartl and A. G. Clark, *Principles of Population Genetics* (2d ed., 1989, Sinauer, Sunderland, MA) is the best that I have found for teaching purposes. More specialized, but still quite useful, is *Introduction to Quantitative Genetics* (D. S. Falconer and T.F.C. Mackay, 4th ed., 1996, Longman, Harlow, Essex). The edited volume *Adaptation* (M. R. Rose and G. V. Lauder, eds., 1996, Academic Press, San Diego) offers an up-to-date assembly of articles on one of the main themes of Darwinism. However, it is hard to recommend these last books to the beginner.

Notes

CHAPTER 1: DARWIN

1. For the life of David Hume, see E. C. Mossner, *The Life of David Hume* (1954, University of Texas Press, Austin).

2. Three examples of eighteenth-century Continental literature are Voltaire's contributions to the great French Encyclopedia as well as his *Candide* (published in 1759), and Giacomo Casanova's *History of My Life* (although only W. R. Trask's translation, published by Johns Hopkins University Press, paperback edition 1997, is an uncorrupted English version).

3. Mary Shelley's book, *Frankenstein, or, A Modern Prometheus*, was first published in 1818, when she was in her early twenties.

4. My main source for the worldwide historical background to the young Charles Darwin's life is Paul Johnson's *The Birth of the Modern World Society, 1815–1830* (1991, HarperCollins, New York).

5. *Sense and Sensibility* was published in 1811. The quotation is from chapter 3.

6. The Darwin quotation is from page 57 of the *Autobiography* edition referenced above.

7. The title of the *Beagle* book is *Journal of Researches into the Natural History and Geology of Countries Visited by H.M.S. Beagle.*

8. The Malthus publication was *An essay on the principle of population*, first appearing in 1798.

9. A. S. Byatt's novella *Morpho Eugenia* appeared in *Angels and Insects*, published by Chatto & Windus, London, in 1992.

CHAPTER 2: HEREDITY

1. The two quotations from the *Origin* are from its chapter I.

2. The article by Fleeming Jenkin was published in 1867 as "The origin of species," *North British Review* 45: 277–318.

3. The Bacon quotation is from *Historia Vitae et Mortis,* pp. 258–59, translated by J. Spedding, R. L. Ellis and D. D. Heath, 1889, Longman, London.

4. Galton's *Hereditary Genius* was first published in 1869 by Macmillan, London.

5. Mendel's first publication on the future genetics was "Versuche über Pflanzen-hybriden," in *Verhandlungen des naturforschenden Vereines in Brünn* 4: 3–47.

6. James D. Watson's great moment of candor came in *The Double Helix: A Personal Account of the Discovery of the Structure of DNA,* published in 1968 by Atheneum, New York.

7. The story about the clash between Pearson and Bateson comes from R. C. Punnett, in a 1950 article, "Early days of genetics," published in *Heredity* 4: 1–10.

8. Fisher's first publication of his synthesis of biometrics and genetics was in 1918, by the title "The correlation between relatives on the supposition of Mendelian inheritance," in *Transactions of the Royal Society of Edinburgh* 52: 399–433.

CHAPTER 3: SELECTION

1. Empedocles is the foremost classical proponent of the theory that life had a material, rather than divine, origin, while Lucretius is regarded by some scholars as the greatest classical forerunner to modern science. Denis Diderot was the principal editor of the great French Encyclopedia of the Enlightenment. Jean-Jacques Rousseau was perhaps the most charismatic of the French-language intellectuals of the Enlightenment, as well as a self-confessed scoundrel. Jean-Jacques Rousseau immortalized his life and character in his *Confessions,* published in 1781. David Hume we have already mentioned in chapter 1.

2. Most of the material on Darwin's approach to the difficulties for the theory of evolution by natural selection is taken from chapter VI of the *Origin.*

3. The best presentation of the resolution of the struggle between Darwinism and Mendelism is that given by Provine, in his aforementioned book.

4. The Jennings quotation is taken from his "Modifying factors and multiple allelomorphs in relation to the results of selection," published in *American Naturalist* 51: 301–6.

5. The key Weldon paper on his crab work is "Attempt to measure the death-rate due to the selective destruction of *Carcinus moenas* with respect to a particular dimension," 1895, *Proceedings of the Royal Society* 58: 360–79.

6. A good summary of butterfly and moth work on industrial melanism is provided by E. B. Ford's *Ecological Genetics* (3d ed. 1971, Chapman and Hall, London).

7. The classic Hamilton papers on kin selection were "The genetical evolution of social behaviour, I & II" (*Journal of Theoretical Biology* 7: 1–52).

8. The definitive book on evolutionary game theory is *Evolution and the Theory of Games* (J. Maynard Smith, 1982, Cambridge University Press, Cambridge, UK).

9. The example of alloparental behavior comes from G. C. Williams's *Adaptation and Natural Selection*.

CHAPTER 4: EVOLUTION

1. Depew and Wagner's *Darwinism Evolving* gives an introduction to classical ideas of taxonomy, which can be read along with Mayr's *Growth of Biological Thought*.

2. The key work by Charles Lyell was *Principles of Geology, Being an Attempt to Explain the Former Changes of the Earth's Surface, by Reference to Causes Now in Operation* (in 3 vols., 1830–33, John Murray, London).

3. The arguments for Darwin's Tree of Life are taken primarily from chapter X of the *Origin*.

4. The classic references to species formation and the nature of reproductive barriers are Dobzhansky's *Genetics and the Origin of Species* and Mayr's *Systematics and the Origin of Species*, already mentioned. These books were updated by Dobzhansky's *Genetics of the Evolutionary Process* (1970, Columbia University Press) and Mayr's *Animal Species and Evolution* (1963, Belknap Press, Cambridge, MA).

5. The original publication of the punctuated equilibrium model was N. Eldredge and S. J. Gould, "Punctuated equilibria: An alternative to phy-

letic gradualism" (1972, pp. 82–115, in T.J.M. Schopf, ed., *Models in Paleobiology*, W. H. Freeman, San Francisco).

6. The story of mass extinctions and large-body impacts has been told for a popular audience by D. M. Raup (1986, *The Nemesis Affair*, W. W. Norton, New York) and W. Alvarez (1997, *T. Rex and the Crater of Doom*, Princeton University Press, Princeton, NJ).

General References and Readings for Part Two

Unlike Part One, the material of Part Two does not naturally come together at the level of reference books. Thus, I will give some general reference books, chapter by chapter, and then proceed to give the individual notes for each chapter.

My main source for the statistical and historical background to the chapter on agriculture is Hugh Thomas (1979, *A History of the World*, HarperCollins, New York), who tells world history in terms of the things that really matter: grain, steam engines, and printing presses. The best introduction to Darwinian breeding is the book by Falconer and Mackay, already mentioned in the bibliography for Part One. More advanced selections of topics in this area are supplied by B. S. Weir, E. J. Eisen, M. M. Goodman, and G. Namkoong, eds., *Proceedings of the Second International Conference on Quantitative Genetics* (1988, Sinauer Associates, Sunderland, NJ); W. G. Hill and T.F.C. Mackay, eds., *Evolution and Animal Breeding* (1989, CAB International, Wallingford, Oxon); and A. A. Hoffman and P. A. Parsons, *Evolutionary Genetics and Environmental Stress* (1991, Oxford University Press, Oxford). As this list indicates, I haven't found any really good introductions to this subject for the general reader.

Darwinian medicine has recently been a fairly hot area. There is a very readable general introduction to this topic from R. M. Nesse and G. C. Williams, *Why We Get Sick* (1994, Random House, New York). As a one-stop survey, it is the next step after reading chapter 6. The next way station along the route would be *Why We Age: What Science Is Discovering about the Body's Journey through Life* (Steven N. Austad, 1997, John Wiley, New York). Beyond this book, the material gets harder fast. Still readable, though requiring some background in biology, is Paul W. Ewald's book on infectious disease, *Evolution of Infectious Disease* (1994, Oxford University Press, New York). Of about the same level of difficulty is *Longevity, Senescence, and the Genome* (C. E. Finch, 1990, University of Chicago Press, Chicago). My own

Evolutionary Biology of Aging (M. R. Rose, 1991, Oxford University Press, New York) can't be recommended to the general reader, though biologists should find it useful. Still farther out there, but outstanding, is Brian Charlesworth's *Evolution in Age-Structured Populations* (2d ed., 1994, Cambridge University Press, Cambridge, UK). Finally, *Genetics and Evolution of Aging* (M. R. Rose and C. E. Finch, eds., 1994, Kluwer, Dodrecht, Netherlands) is close to state of the art.

Unlike the other two topics, eugenics has been gone over and over by the social science book mill, perhaps because the subject is both somewhat lurid and safely dead. There are many fine, articulate, historical books in this area. Among the best is Daniel J. Kevles's *In the Name of Eugenics: Genetics and the Uses of Human Heredity* (2d ed., 1995, Harvard University Press, Cambridge, MA). The Kevles book tends to take an Anglo-American focus, with the full-scale disaster of the Nazis underplayed to my taste. Diane B. Paul provides a recent, general, and concise treatment in her *Controlling Human Heredity, 1865 to the Present* (1995, Humanities Press, New Jersey). Stefan Kühl (*The Nazi Connection: Eugenics, American Racism, and German National Socialism*, 1994, Oxford University Press, New York) does a good job of bringing out the ugliness on an international scale, focusing on the many connections between eugenics and racism. More specific treatment of racism is supplied by Pat Shipman (*The Evolution of Racism, Human Differences and the Use and Abuse of Science*, 1994, Simon & Schuster, New York). A bridge connecting eugenics and racism to recent applications of genetics is built by R. G. Steen (*DNA and Destiny: Nature and Nurture in Human Behavior*, 1996, Plenum, New York). These books also reference the older, almost endless, literature on eugenics. One outstanding book on the actual complexity of differentiation between human populations is that of L. L. Cavilli-Sforza, P. Menozzi, and A. Piazza, *The History and Geography of Human Genes* (1994, Princeton University Press, Princeton, NJ). A serious reading of this book, or its equivalent, should disabuse any objective reader of the salience of the race concept applied to human biology.

Notes

CHAPTER 5: AGRICULTURE

1. Among the leading practitioners of Darwinism that have worked within an agricultural context are Oscar Kempthorne, Alan Robertson, and

C. Clark Cockerham. See the volumes edited by Weir et al. and Hill and Mackay, above.

2. The historical details about agricultural production come from the Thomas world history.

3. The Spanish pointer is discussed in chapter I of the *Origin*.

4. Sewall Wright's 1952 article on quantitative genetics was "The genetics of quantitative variability," pp. 5–41 in *Quantitative Inheritance* (Agricultural Research Council, H.M.S.O., London). For more on Wright's work, see *Sewall Wright and Evolutionary Biology* (W. B. Provine, 1986, University of Chicago Press, Chicago).

5. The *Evolution and Animal Breeding* book is that of Hill and Mackay, mentioned earlier.

6. Again, see Provine's 1986 book for more on Sewall Wright.

7. The final quotation is taken from p. 402 of Thomas.

CHAPTER 6: MEDICINE

1. References to Nesse and Williams refer to their book, *Why We Get Sick*, mentioned above.

2. Finch (1990) has a good discussion of marsupial mice, as well as reviewing material on the effects of castration.

3. *Larousse Encyclopedia of Mythology* (ed. by Robert Graves, republished 1994, Barnes & Noble Books, New York).

4. The data on human longevity come from D.W.E. Smith, *Human Longevity* (1993, Oxford University Press, New York).

5. The book by Finch (*Longevity, Senescence, and the Genome*) is a source of much useful information on castration and sexual function, generally.

6. One manual listing human genetic disorders is that of V. A. McKusick and Clair A. Francomano, *Mendelian Inheritance in Man: A Catalog of Human Genes and Genetic Disorders* (11th ed., 1994, Johns Hopkins University Press, Baltimore).

7. The calculations on the impact of incest are based on information from J. F. Crow and M. Kimura, *An Introduction to Population Genetics Theory* (1970, Harper & Row, New York).

8. Ewald (1994) really pioneered the evolutionary biology of disease defenses, and all discussions of the problem, including mine here, follow him.

9. Frank (1996) provides a nice introduction to the immune system as internal Darwinism, in Rose and Lauder (1996).

10. See Finch (1990) and Rose and Finch (1994) for more on Huntington's disease.

11. The experiments on laboratory evolution of postponed aging are covered in chapter 3 of *Evolutionary Biology of Aging* (Rose, 1991).

CHAPTER 7: EUGENICS

1. The opening mythology is also taken from the *Larousse Encyclopedia of Mythology.*

2. Galton's first eugenics proclamations came in "Hereditary talent and character, I & II," published in 1865 in *Macmillan's Magazine* 12: no. 68, pp. 157–66 and no. 71, pp. 318–27. See other original eugenics publications in *Eugenics, Then and Now* (C. J. Bajema, 1976, Dowden, Hutchinson, & Ross, Stroudsburg, PA). For a laudatory biography of Galton, there is *Eugenics, Galton and After* (C. P. Blacker, 1952, Duckworth, London).

3. The Punnet reference is to "Eliminating feeblemindedness," *Journal of Heredity* 8: 464–65.

4. Kevles (1995) provides the most focused account of eugenics in the USA, especially the machinations of Charles Davenport.

5. The Futuyma book is the same as the one referred to earlier, pp. 107–9.

6. See Kühl and Steen, above, on the Nazi health courts.

7. The publication data for the Jennings book are Dutton & Co., New York.

General References and Readings for Part Three

The topic of human nature can only be considered a focus for warring opinions, not an area of settled knowledge, at least not with respect to the selection pressures, and their adaptive consequences, responsible for human evolution. Any book that has anything interesting to say in this area is only offering an opinion. The book that created the modern era in human evolutionary biology is E. O. Wilson's *Sociobiology, The New Synthesis* (1975, Belknap Press, Cambridge, MA). He then followed this book up with *On Human Nature* (1978, Harvard University Press, Cambridge, MA); *Genes, Mind, and Culture* (C. Lumsden and E. O. Wilson, 1981, Harvard University Press, Cambridge, MA); and *Promethean Fire* (C. Lumsden and E. O. Wilson, 1983, Harvard University Press, Cambridge, MA). Another book developing the possibilities for interaction between culture and biological evolution is *Culture and Evolutionary Process* (R. Boyd and P. J. Richerson, 1985, University of Chicago Press, Chicago).

The fallout from sociobiology has been extensive. Here are just some of the declamations that made it into print. *The Use and Abuse of Biology: An Anthropological Critique of Sociobiology* (M. Sahlins, 1976, University of Michigan Press, Ann Arbor). The articles in *The Sociobiology Debate* (A. L. Caplan, ed., 1978, Harper, New York) are both pro and con. *Morality as a Biological Phenomenon: The Presuppositions of Sociobiological Research* (G. S. Stent, ed., rev. ed., 1980, University of California Press, Berkeley). *Sociobiology Examined* (A. Montagu, ed., 1980, Oxford University Press, New York). *Vaulting Ambition, Sociobiology and the Quest for Human Nature* (P. Kitcher, 1985, MIT Press, Cambridge, MA).

The evolutionary psychology movement is founded primarily on two books: Jerome H. Barkow's *Darwin, Sex, and Status: Biological Approaches to Mind and Culture* (1989, University of Toronto Press, Toronto) and J. H. Barkow, L. Cosmides, and J. Tooby, eds., *The Adapted Mind, Evolutionary Psychology and the Generation of Culture* (1992, Oxford University Press, New York). Transitional figures between sociobiology and evolutionary psychology might include R. D. Alexander (e.g., *Darwinism and Human Affairs*, 1979, University of Washington Press, Seattle). But there are numerous other publications that owe their allegiance to the flag of sociobiology and evolutionary psychology. Some of these have found their way into *Human Nature: A Critical Reader* (L. Betzig, ed., 1997, Oxford University Press, New York), which may be the best value for money among introductions to evolutionary psychology. A specifically popular account of evolutionary psychology is *The Moral Animal: Evolutionary Psychology and Everyday Life* (Robert Wright, 1994, Random House, New York).

The topic of religion is perhaps one of the primary sources of profit for publishing houses. To this day, Oxford University Press makes more money off the Bible than any other publication. One of the publications about the historical relationship between Darwinism and religion that I have found most useful is that of Ron Amundson, "Historical development of the concept of adaptation," in Rose and Lauder (1996), described above. Depew and Weber (1995), a general reference for Part One, also provide an interesting long-range view.

Notes

CHAPTER 8: ORIGINS

1. For an example of a "savanna story," see "The origin of man" (C. O. Lovejoy, 1981, *Science* 211: 341–50).

2. A recent review of the human paleontology literature is *Paleoanthropology* (M. H. Wolpoff, 1997, McGraw-Hill, New York).

3. The central figure advocating the neutral theory of evolution was M. Kimura (e.g., *The Neutral Theory of Molecular Evolution,* 1983, Cambridge University Press, Cambridge).

4. Even Stephen Jay Gould, who avoids the invocation of natural selection in human evolution, concedes the difficulties facing human pregnancy and birth (e.g., "Human Babies as Embryos," pp. 70–75 in *Ever Since Darwin,* 1978, Burnett, London).

5. Engels's essay was published in his *Dialectics of Nature.*

6. The Oakley reference is K. Oakley, *Man the Tool-Maker* (1959, University of Chicago Press, Chicago).

7. For one example of Washburn's views, see S. L. Washburn, "Tools and human evolution," 1960, *Scientific American* 203: 62–75.

8. The classic social intelligence reference is N. K. Humphrey, "The function of intellect," in *Growing Points in Ethology* (P.P.G. Bateson and R. A. Hinde, eds., 1976, Cambridge University Press, Cambridge, UK, pp. 303–17). See also *Machiavellian Intelligence, Social Expertise and the Evolution of Intellect in Monkeys, Apes, and Humans* (R. Byrne and A. Whiten, eds., 1988, Clarendon Press, Oxford).

9. Some examples of arms-race evolution publications include *The Dawn Warriors: Man's Evolution toward Peace* (R. S. Bigelow, 1969, Little Brown, Boston) and the Alexander book mentioned above.

10. The discussion of the formal features of mental arms races is drawn from my 1980 article, "The mental arms race amplifier," *Human Ecology* 8: 285–93.

CHAPTER 9: PSYCHE

1. The range of debate among zoologists and evolutionary biologists is illustrated well by the book edited by Kaplan (1978).

2. An example of a dismissive *New York Review of Books* article is the Stephen Jay Gould review of the Lumsden and Wilson book, *Promethean Fire,* from June 30, 1983. The title on the cover is "Sociobiology, Goodbye."

3. The full citation for the spandrels article is S. J. Gould and R. C. Lewontin, 1979, "The spandrels of San Marco and the Panglossian paradigm. A critique of the adaptationist program," *Proceedings of the Royal Society B* 205: 581–98. The Rose and Lauder book *Adaptation* is in large part a response to this article and its effects on a generation of evolutionary biologists.

4. Alexander (1979) provides a good review of the incest avoidance phenomenon.

5. Sahlins (1976) was the first to bring forward the kinship problem for sociobiology.

6. Lumsden and Wilson citations are given above.

7. Likewise, see the bibliography for this Part for the Barkow books.

8. An example of evolutionary psychology approaching sexual selection is *The Evolution of Desire: Strategies of Human Mating* (David M. Buss, 1994, HarperCollins, New York).

9. Examples of research on murder and other violent acts can be found in Betzig's book, above.

10. Examples of the application of Darwinism to animal behavior as the stuff of behavioral ecology are given by *An Introduction to Behavioral Ecology* (J. R. Krebs and N. B. Davies, 1981, Blackwell, Oxford).

11. The criminology of sociopaths is described in *Crime and Human Nature*, (James Q. Wilson and R. J. Herrnstein, 1985, Simon & Schuster, New York).

12. The clinical and general psychology of sociopaths is described by Hervey Cleckley's *The Mask of Sanity* (5th ed., 1988, A. Cleckley, Augusta, GA).

13. See L. Mealey, 1995, "The sociobiology of sociopathy: An integrated evolutionary model," *Behavioral and Brain Sciences* 18: 523–99.

14. This analysis was developed by Chris Moore and myself in M. R. Rose and C. Moore, 1993, "A Darwinian function for the orbital cortex," *Journal of Theoretical Biology* 161: 119–29.

15. A popular introduction to the literature on frontal lobe and related brain disorders is provided by Antonio R. Damasio, *Descartes' Error: Emotion, Reason, and the Human Brain* (1994, Putnam Berkley, New York).

CHAPTER 10: SOCIETY

1. A general introduction to the history of economics is William J. Barber, *A History of Economic Thought* (1967, Penguin, Harmondsworth, UK), but there are many other books of this kind.

2. John Maynard Keynes, *The collected writings of John Maynard Keynes* (1971–89, Macmillan, London).

3. Some of the writings of the neo-Austrians are *Human Action: A Treatise on Economics* (4th ed., Ludwig von Mises, 1996, Foundation for Economic Education, Irvington-on-Hudson, NY); *The Collected Works of Friedrich August Hayek* (W. W. Bartley III, ed., 1988–95, Routledge, London); and *Capi-*

talism, Socialism, and Democracy (Joseph A. Schumpeter, 1942, Harper, New York).

4. For an expanded version of this argument, see my "Hominid evolution and social science," 1983, *Journal of Social and Biological Structures* 6: 29–36.

5. See the introduction to Michael Oakeshott, ed., *Leviathan*, by Thomas Hobbes (1946, Blackwell, London).

6. See Depew and Weber, above.

7. Described in Browne's *Voyaging*, above.

8. The intellectual history of Social Darwinism is very complex. For one introduction, see Robert C. Bannister, *Social Darwinism, Science and Myth in Anglo-American Thought* (1979, Temple University Press, Philadelphia).

9. Published in 1974 by Basic Books, New York.

10. See, for example, Jean-Jacques Rousseau, *The Social Contract and the Discourses* (G.D.H. Cole, trans., J. H. Brumfitt and J. C. Hall, eds., 1993, A. A. Knopf, New York.)

11. See *The Complete Works of George Orwell* (1986, Secker & Warburg, London).

12. Daniel Bell, *The End of Ideology: On the exhaustion of political ideas in the fifties* (1960, Free Press, Glencoe, IL).

13. Francis Fukuyama, *Have we reached the end of history?* (1989, Rand Corporation, Santa Monica, CA).

14. Lyrics from the song "Spirits in the Material World," from The Police's 1981 LP *Ghost in the Machine*, A&M Records.

CHAPTER 11: RELIGION

1. For the Darwinian side of the story, see Douglas J. Futuyma, *Science on Trial: The Case for Evolution* (1995, Sinauer, Sunderland, MA).

2. See Stillman Drake, *Gallileo at Work: His Scientific Biography* (1978, University of Chicago Press, Chicago).

3. See the chapter by Amundson in Rose and Lauder (1996).

4. The most sophisticated advocate for scientific creationism is Phillip E. Johnson (e.g., *Darwin on Trial*, 1991, Regnery Gateway, Lanham, MD), a law professor at the University of California, Berkeley.

5. The classic compendium of religious experiences is William James's *The Varieties of Religious Experience* (first published in 1902, and since republished in many editions).

6. Walter Burkert, *Creation of the Sacred: Tracks of Biology in Early Religions* (1996, Harvard University Press, Cambridge, MA) offers another analysis of the cultural roles of religion, and the advantages it affords the faithful.

7. Julian Jaynes, *The Origin of Consciousness in the Breakdown of the Bicameral Mind* (1976, Houghton Mifflin, Boston).

INDEX